Thermo-Hydro-Mechanical Coupling in Clay Rock

Weizhong Chen • Hongdan Yu • Xiangling Li

Thermo-Hydro-Mechanical Coupling in Clay Rock

Experimental Characterization and Numerical Modelling

Springer

Weizhong Chen
Institute of Rock and Soil Mechanics
Chinese Academy of Sciences
Wuhan, China

Hongdan Yu
Institute of Rock and Soil Mechanics
Chinese Academy of Sciences
Wuhan, China

Xiangling Li
Tongji University
Shanghai, China

ISBN 978-981-95-2561-4 ISBN 978-981-95-2562-1 (eBook)
https://doi.org/10.1007/978-981-95-2562-1

Jointly published with Science Press

The print edition is not for sale in China (Mainland). Customers from China (Mainland) please order the print book from: Science Press.

This Springer imprint is published by the registered company Springer Nature Singapore Pte Ltd.
The registered company address is: 152 Beach Road, #21-01/04 Gateway East, Singapore 189721, Singapore

Preface

The management of spent nuclear fuel and long-lived radioactive waste constitutes a critical challenge for nuclear-powered nations, intersecting nuclear sustainability, industrial safety, and planetary environmental stewardship. Safe geological disposal of high-level nuclear waste (HLW) represents not only a technical imperative but also a cornerstone of intergenerational ecological stewardship. Among disposal strategies, deep geological repositories (DGRs) have emerged as the globally recognized optimal solution for HLW isolation. Within this context, argillaceous formations—exemplified by Boom Clay (HADES, Belgium), Callovo-Oxfordian claystone (Meuse/Haute-Marne, France), and Opalinus clay (Mont Terri, Switzerland)—demonstrate superiority through their low hydraulic conductivity, high cation exchange capacity, and pressure-induced sealing mechanisms, making them preferred host rocks for DGR implementations.

This monograph presents three principal contributions:

- Systematic documentation of thermo-hydro-mechanical (THM) experimental results from natural clay specimens, integrating mineralogical characterization, microstructural analysis, and advanced mechanical testing protocols.
- Development of coupled elasto-viscoplastic constitutive frameworks, incorporating time-dependent behavior and damage evolution mechanics, with a particular emphasis on clayey media self-sealing dynamics under confinement.
- The development of a multi-scale numerical modeling system enables characterization of spatiotemporal thermo-hydro-mechanical (THM) coupling phenomena during repository construction (characterized by stress redistribution dynamics from connecting gallery excavation) and operational phases (demonstrated through PRACLAY large-scale in situ heating tests), with model validity systematically verified against comprehensive in situ monitoring datasets from the CLIPEX project.

The authors extend their sincere appreciation to EURIDICE (European Underground Research Infrastructure for Disposal of Nuclear Waste in Clay Environment) for their indispensable contributions to the China–Belgium decade-long collaborative research program on the thermo-hydro-mechanical (THM) behavior of Boom Clay.

The strategic partnership, established in 2007 and continuously reinforced through joint experimental campaigns and data-sharing protocols, has fundamentally advanced our understanding of clay rock deformation mechanisms under coupled thermo-hydro-mechanical stresses. The National Natural Science Foundation of China (NSFC Grants 42377179, 42293355, and 51879258) and EURIDICE (Grants EUR-07-206 and EUR-12-110) are jointly acknowledged for their synergistic financial support, which has underpinned the program's continuity and scientific impact.

Wuhan, China Weizhong Chen
Wuhan, China Hongdan Yu
Mol, Belgium Xiangling Li

Competing Interests The authors have no competing interests to declare that are relevant to the content of this manuscript.

Declaration of Competing Interest

The content herein reflects solely the authors' research achivements and perspectives, it does not represent the official positions of their affiliated institutions or any third-party entities.

Contents

Chapter 1
Introduction

Thermo-hydro-mechanical (THM) coupling phenomena manifest in subsurface engineering systems spanning enhanced geothermal systems (EGS), coalbed methane extraction, high-level nuclear waste repositories, carbon dioxide sequestration formations, and enhanced oil recovery (EOR) operations within petroleum reservoirs.

The scientific exploration of thermo-hydro-mechanical (THM) coupling in clayey formations originated from Bouyouco's (1915) foundational investigation into thermal sensitivity of soils, evolving into a critical focus for high-level radioactive waste (HLW) geological disposal. This paradigm shift has driven interdisciplinary research characterizing coupled thermal-hydro-mechanical-chemical (THMC) interactions through experimental quantification of heat-fluid-rock feedback mechanisms. Landmark in situ experiments in European clay host rocks (Belgium's HADES, France's Meuse/Haute-Marne, and Switzerland's Mont Terri) have elucidated coupled THM processes under repository-relevant confinement conditions (Bossart et al. 2004; Zhang 2013; Chen et al. 2011, 2024). The DECOVALEX international consortium, established in 1992 and now in its ninth phase (2024–2027), serves as a paradigm for interdisciplinary synergy, integrating geotechnical engineers, hydrologists, and nuclear safety regulators to develop predictive models for THMC coupling. Despite progress in experimental characterization, the inherent complexity of coupled processes—where thermal diffusion (μs-ms), fluid flow (days-years), and mechanical/chemical relaxation (years-centuries) exhibit divergent temporal responses—poses fundamental challenges for developing unified modeling frameworks that resolve multi-physics interactions across material-specific activation timescales (Fig. 1.1).

W. Chen et al., *Thermo-Hydro-Mechanical Coupling in Clay Rock*,
https://doi.org/10.1007/978-981-95-2562-1_1

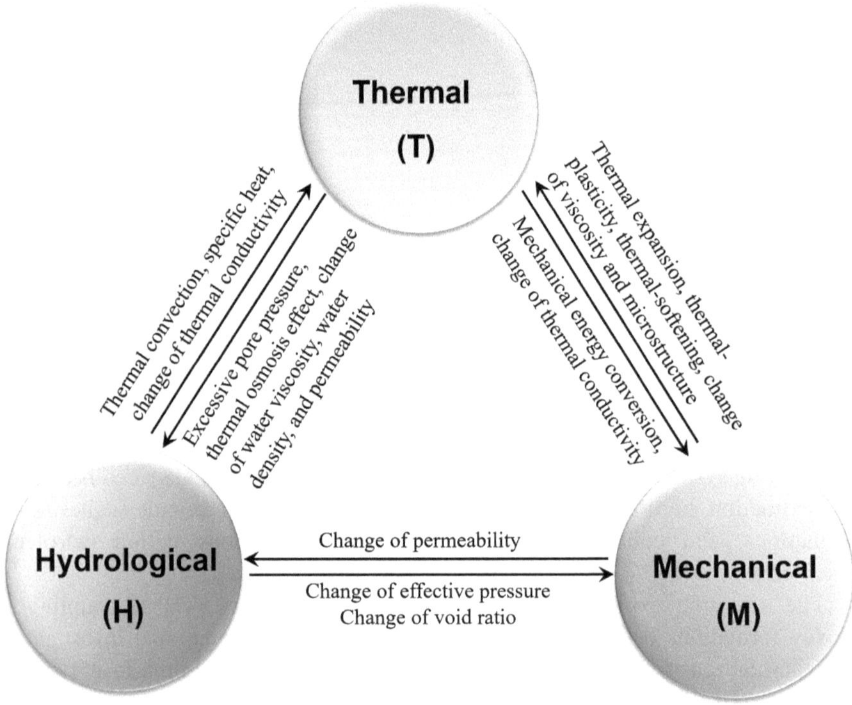

Fig. 1.1 Coupled THM processes in clays

1.1 Temperature Effects on the Mechanical Properties of Clay Rocks

1.1.1 Temperature Effects on the Strength and Stiffness of Clay Rocks

The thermal effects on the stiffness and strength of clay rock have always been controversial. Field investigations have been conducted into this subject (Chen et al. 2015). Isothermal mechanical testing is a commonly used method in which the sample is usually loaded at a constant temperature. These tests are usually conducted with temperature-controlled triaxial compression testing devices, which consist of conventional triaxial compression testing devices and heating systems (oven or heating coil). There are many different opinions on the effects of temperature on the mechanical behavior of clay rocks, a summary of which is given in Table 1.1 and is discussed below.

Hueckel and Baldi (1990), Ghahremannejad (2006), De Bruyn and Thimus (1996, Fig. 1.2), Noble (1968), Zhang et al. (2007, 2017), and Yu et al. (2018, 2021a, 2022a) indicated that the strength and elastic modulus of clay rocks decrease with an increase in temperature. However, Abuel-Naga et al. (2006), Chen et al. (2005),

Table 1.1 Review of temperature effects on Young's modulus and shear strength of clay rock

Author	Clay	Temperature range/°C	Consolidation state	Test method	Young's modulus	Peak strength	Slope of CSL
Hueckel and Baldi (1990)	Pontida silty clay	23–98	OC	CU	↘	↘	→
De Bruyn and Thimus (1996)	Boom Clay	20–80	OC	CU	–	↘	→
Kuntiwattanakul et al. (1995)	Kaolin clay	20–90	OC	CU	→	→	–
			NC		↗	↗	–
Burghignoli et al. (2000)	Todi clay	15–60	NC	CU	→	→	→
Chen et al. (2005)	Nanyang expansive clay	15–60	NC	CD	↗	↗	↗
Ghahremannejad (2006)	M44 clay	22–100	NC	CD	↘	↘	–
Tanaka et al. (1997)	Illitic clay	28–100	NC, OC	CD	–		↘
Cekerevac and Laloui (2004)	Kaolin clay	22–90	NC, OC	CD	↗	↗	→
Abuel-Naga et al. (2006)	Bangkok clay	25–90	NC	CU	–	↗	→
Zhang et al. (2007, 2017)	Opalinus clay	20–116	–	CU	–	↘	–
Schuster et al. (2021)	Opalinus clay	25–200	–	UU	–	→	–
Yu et al. (2018)	Boom Clay	22–80	OC	CU	↘	↘	↗
Menaceur et al. (2015)	COx claystone	25–80	–	CD	→	↘	–
Liu et al. (2019)	COx claystone	20–90	–	CD	→	↘	–
Gbewade et al. (2024)	COx claystone	20–100	–	CD	→	↘	–
Gbewade et al. (2024)	COx claystone	100–150	–	CD	↗	↗	–
Yu et al. (2024a)	Tamusu mudstone	22–80	–	CD	↗	↗	–

[a]*OC* overconsolidated, *NC* normally consolidated, *CU* consolidated undrained, *CD* consolidated drained, *UU* unconfined undrained, *COx* Callovo-Oxfordian

Cekerevac and Laloui (2004), Menaceur et al. (2015), Liu et al. (2019), and Yu et al. (2024a) obtained contradictory conclusions. What's more, Schuster et al. (2021) indicated that the total strain at failure, Young's moduli, and brittleness values of Opalinus clay are independent of temperature (Fig. 1.3), which is similar to the observation of Burghignoli et al. (2000). While Gbewade et al. (2024) found that the strength variation law of Callovo-Oxfordian (COx) claystone has changed under

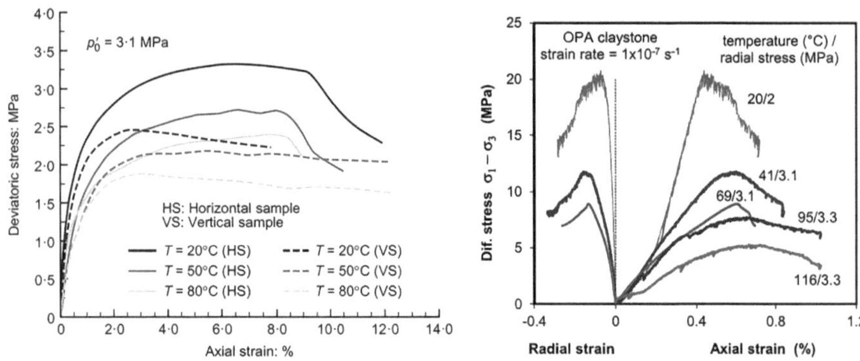

Fig. 1.2 Triaxial undrained compression behavior of (**a**) Boom Clay (De Bruyn and Thimus 1996) and (**b**) Opalinus clay (Zhang et al. 2017) at different temperatures

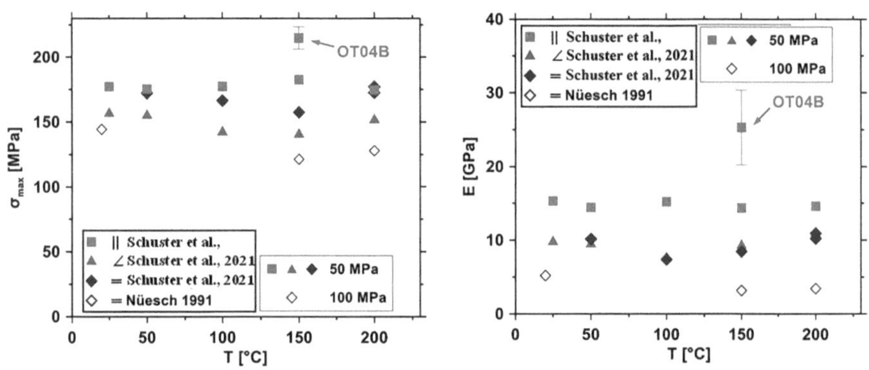

Fig. 1.3 The relation between temperature T and (**a**) peak strength σ_{max} and (**b**) Young's modulus E of Opalinus clay (Schuster et al. 2021; Nüesch 1991)

different temperature conditions, it decreased with temperature increase at the temperature range of 0–100 °C, but increased at the temperature range of 100–150 °C (Fig. 1.4).

Kuntiwattanakul et al. (1995) focuses on another impact aspect, i.e., consolidated state. He conducted CU triaxial compression tests on the normally consolidated and overconsolidated kaolin clay at 20 and 90 °C. He observed that the thermal effects on the strength of clay were dependent on the overconsolidation ratio: the undrained strength of normally consolidated clay was independent of temperature, whereas that of consolidated clay increased with an increase in temperature. Noble (1968) has elaborated on the effects of thermo-mechanical loading paths on clay strength. As indicated in Fig. 1.5, two thermo-mechanical loading paths before shear (Path 1, first heating and then consolidating, and Path 2, first consolidating and then heating) were taken in his tests. The results clearly show that if the clay is consolidated to the same pressure and sheared at the same temperature, an

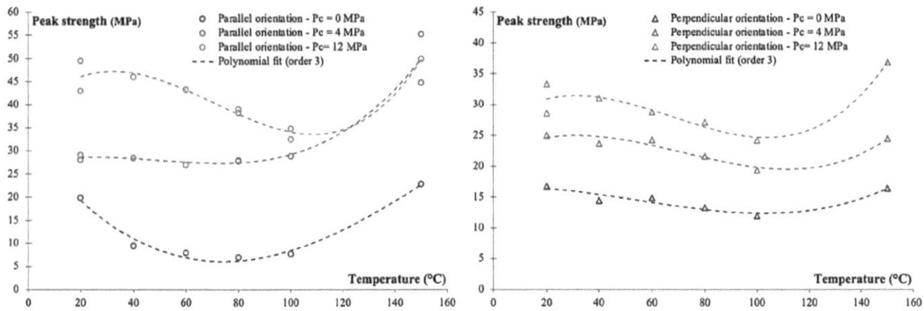

Fig. 1.4 Evolution of the deviatoric stress at peak (peak strength) as a function of temperature for parallel (left) and perpendicular (right) orientations and for all confining pressures (0, 4, and 12 MPa), with polynomial (third order) fitting curves of COx claystone (Gbewade et al. 2024)

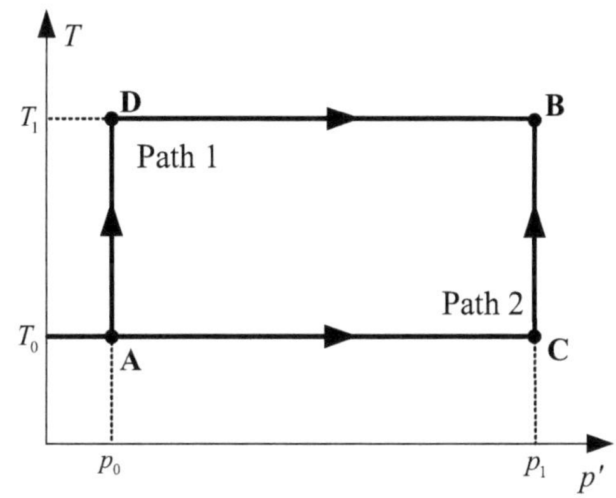

Fig. 1.5 Different thermo-mechanical loading paths before triaxial compression (Noble 1968)

increase in temperature can increase the strength of the clay. For the same final state (Point B), the strength of clay samples taking Path 1 is higher than that taking Path 2.

The thermal effect on the slope of the normally consolidated line (NCL) or the critical state line (CSL) has been conflictingly concluded by previous researchers. Generally speaking, considerable research studies have shown that the slope of NCL remained constant with varying temperatures (Cekerevac and Laloui 2004; Hueckel and Baldi 1990; Burghignoli et al. 2000; Abuel-Naga et al. 2006; Cui et al. 2009). This means that the friction angle of clay rocks at the critical state is unaffected by temperature and that the modification of cohesion causes the variation of clay rock strength with temperature. However, exceptions have also been reported, e.g., the work by Tanaka et al. (1997), in which the slope of NCL decreased with an increase in temperature. In contrast, the work by Chen et al. (2005) indicates that the slope of NCL increased with an increase in temperature. Referring to the slope M of CSL, Hueckel and Pellegrini (1992) and Abuel-Naga et al. (2009) found that the M value of clay rock was independent of temperature. Hueckel (1997) conducted a series of experiments on Boom Clay and found that the M value increased with temperature. Later, Hueckel et al. (2009) found that the temperature effect on the M value might be related to the type of clay rock. Specifically, for Boom Clay, the M value might increase by about 20% when the temperature is increased to 70 °C, but for kaolin, the M value increases by about 12.5% when the temperature is increased to 68 °C. The experimental study from Wang et al. (2021) shows that the slope of the critical state line for kaolin is dependent on the temperature history.

1.1.2 Temperature Effects on the Volumetric Strain of Clay Rocks

Temperature changes may induce the thermal volume change of clay rocks. As early as 1968, Campanella and Mitchell (1968) reported that the irreversible volume decreases after cyclic variation of temperatures. The thermal volume change of clay rock with increasing temperature under constant stress is closely related to its degree of consolidation. Demars and Charles (1982) found that the volume decrease was larger for more plastic soils, and smaller irrecoverable strains were observed at larger OCR values, in agreement with the findings of Paaswell (1967). Heating a normally consolidated sample under a constant load may induce overconsolidation (Cui et al. 2000). Overconsolidated clay generally exhibits thermoelastic expansion, while normally consolidated clay exhibits thermoplastic contraction.

For instance, in the case of Boom Clay, Baldi et al. (1988) observed that the temperature at which the transition between contraction and expansion occurred was changing with OCR. This phenomenon was later confirmed by Cui et al. (2000) and Sultan et al. (2002) (Fig. 1.6). Irrecoverable thermal volume change of a

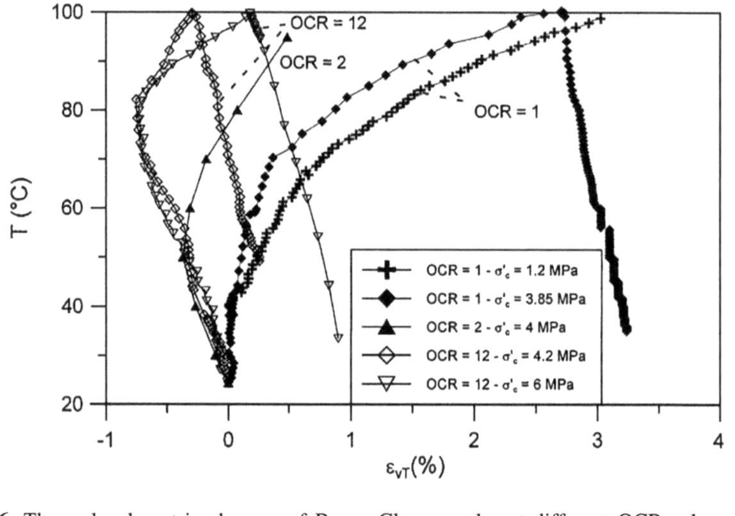

Fig. 1.6 Thermal volumetric changes of Boom Clay samples at different OCR values (Sultan et al. 2002)

normally consolidated clay sample submitted to a constant load is independent of the applied load (Cui et al. 2000; Sultan et al. 2002).

In the case of Opalinus clay, Monfared et al. (2011b) reported a thermal hardening process on the Opalinus Clay, with first a thermoelastic expansion up to 65 °C followed by thermoplastic contraction when heating up to 80 °C, and the memory of the historical largest temperature (65 °C) supported has been erased when heating at 80 °C (Fig. 1.7). Similar results were obtained by Zhang et al. (2017), who observed that indurated clayey materials exhibit an elastic (reversible) thermal expansion, followed by a transition to a plastic thermal compression at some threshold temperature.

In case of COx claystone, Belmokhtar et al. (2017) observed that a thermoelastic expansion occurred up to a temperature of 48 °C, followed by thermoplastic contraction at higher temperatures for COx claystone (Fig. 1.8). This feature was, however, not observed by Mohajerani et al. (2013) and Menaceur et al. (2016) on the COx claystone, who only observed a thermoplastic contraction that could not be suspected from the in situ thermal experiments in which thermoporoelastic models provided satisfactory responses (Gens et al. 2007). Another observation that contradicts Monfared et al. (2011b) is also shown by Belmokhtar et al. (2017). Monfared et al. (2011b) found the memory of the historical largest temperature (65 °C)

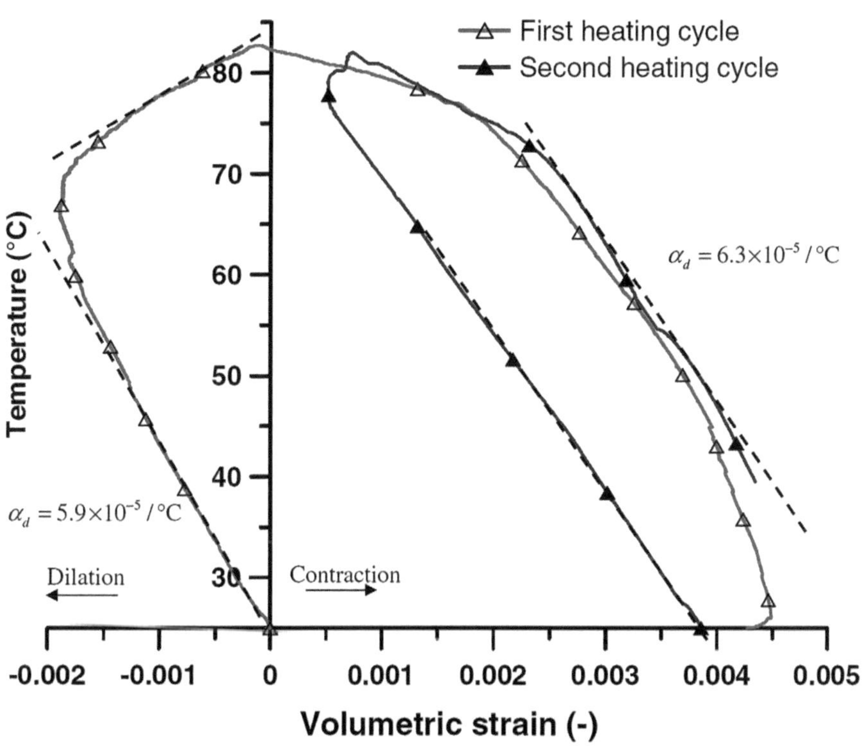

Fig. 1.7 Volumetric response of Opalinus clay during thermal loading (Monfared et al. 2011b)

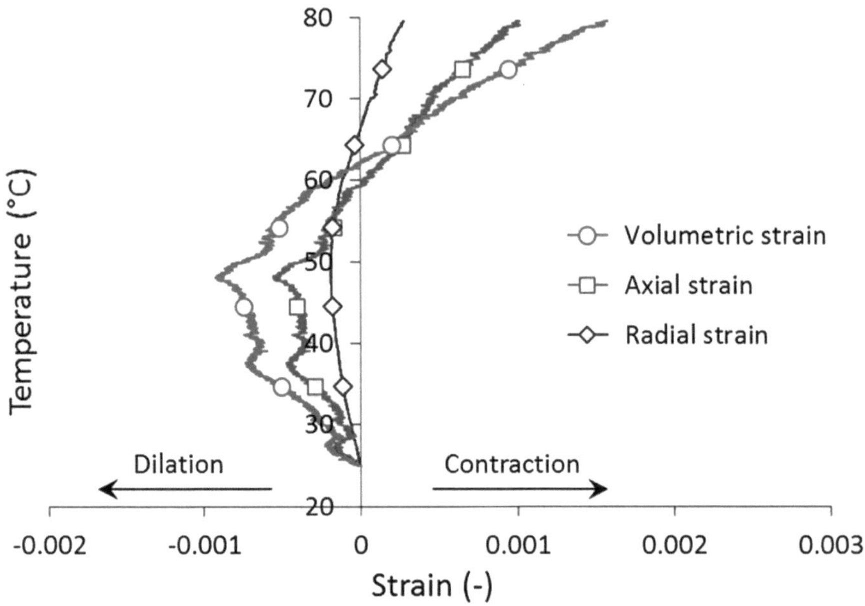

Fig. 1.8 Volumetric response of COx claystone during thermal loading (Belmokhtar et al. 2017)

supported has been erased when heating at 80 °C for Opalinus clay. While Belmokhtar et al. (2017) reported that the memory of the highest experienced temperature was kept by comparing the temperature of 48 °C at which the transition passed from thermoelastic compression to plastic contraction with a maximum burial temperature of 50 ± 5 °C for the COx claystone. If the latter study holds, it means that temperature has an effect comparable to that of stress. This may be compatible with existing thermoplastic models that define an elastic zone and a yield curve in the p, T plane (e.g., Hueckel and Borsetto 1990; Sultan et al. 2002).

1.1.3 Temperature Effect on the Time-Dependent Behavior of Clay Rocks

There are notable differences between different clay formations, reflecting their various origins and geological histories. However, time-dependent behavior is a crucial characteristic among their common key features (e.g. creep, anisotropy, some degree of softening, and variation of permeability with damage). Hence, many scholars were focused on studying the creep deformation laws and induction mechanisms of clay rocks under complex engineering conditions. The creep behavior of clay rocks has been extensively studied by scholars like Arulanandan et al. (1972), Bonini et al. (2003), Gasc-Barbier et al. (2004), Zhang and Rothfuchs (2004), Fabre and Pellet (2006), Naumann et al. (2007), Yu et al. (2015, 2021), Zhang and Laurich

(2020), and Shi et al. (2022), who conducted a series of creep tests on different types of clay rocks, including Boom Clay, COx claystone, Opalinus clay (Fig. 1.9), and Bure clay. All their studies highlight the high creep potential of clay rocks.

In recent years, scholars have gradually begun to pay attention to the creep behavior of clay rocks under high temperatures, and meaningful results have been obtained. However, the existing relevant experimental studies are still very limited and insufficient to fully reveal the creep laws and mechanisms of clay rocks under complex thermo-hydro-mechanical coupling conditions. Indeed, variations in mineral composition, microstructure, and diagenetic conditions result in significant differences in the physical and mechanical responses of rocks to temperature and stress. For instance, Gasc-Barbier et al. (2004), Naumann et al. (2007), Zhang et al. (2007, 2017), Belmokhtar et al. (2017), and Yu et al. (2022a, 2023b) conducted creep tests on Bure Clay, Opalinus clay, and COx claystone (Figs. 1.10 and 1.11), and Boom Clay under different temperature and stress conditions. These studies

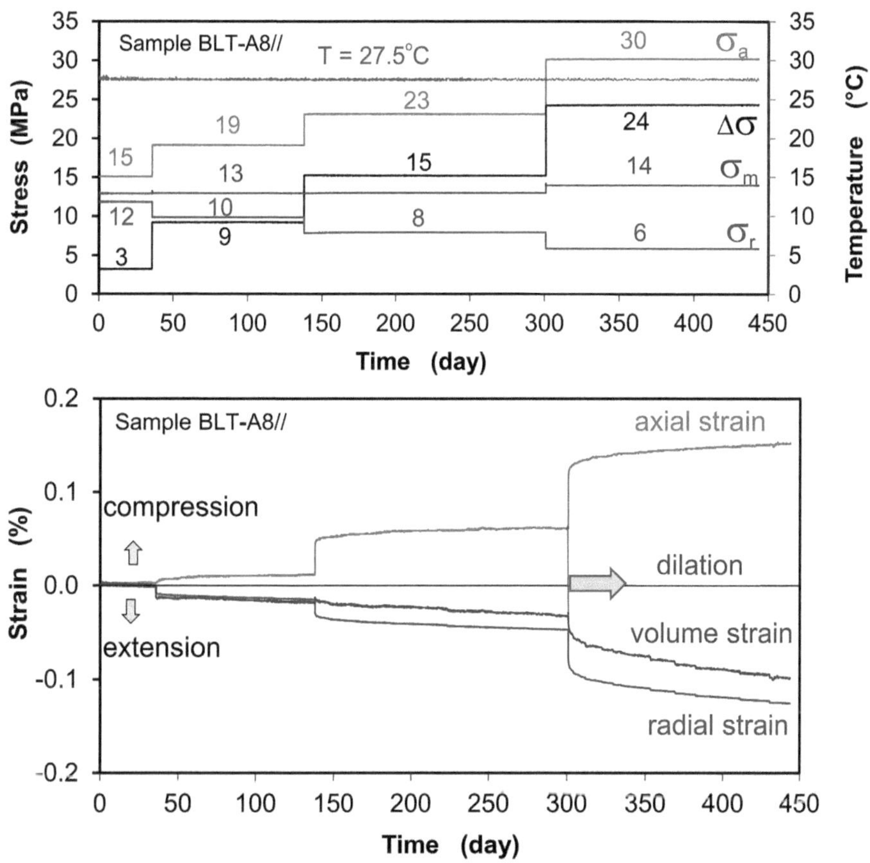

Fig. 1.9 Results of a multistep triaxial creep test on a highly saturated Opalinus clay sample (Zhang and Laurich 2020)

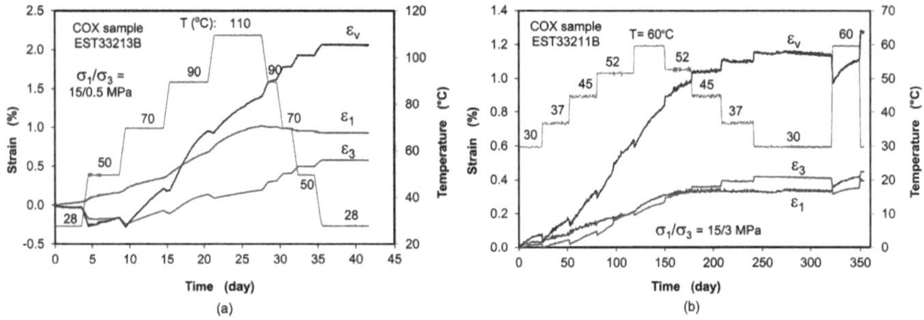

Fig. 1.10 Creep of COx samples during heating/cooling cycle at different temperatures (Zhang et al. 2017)

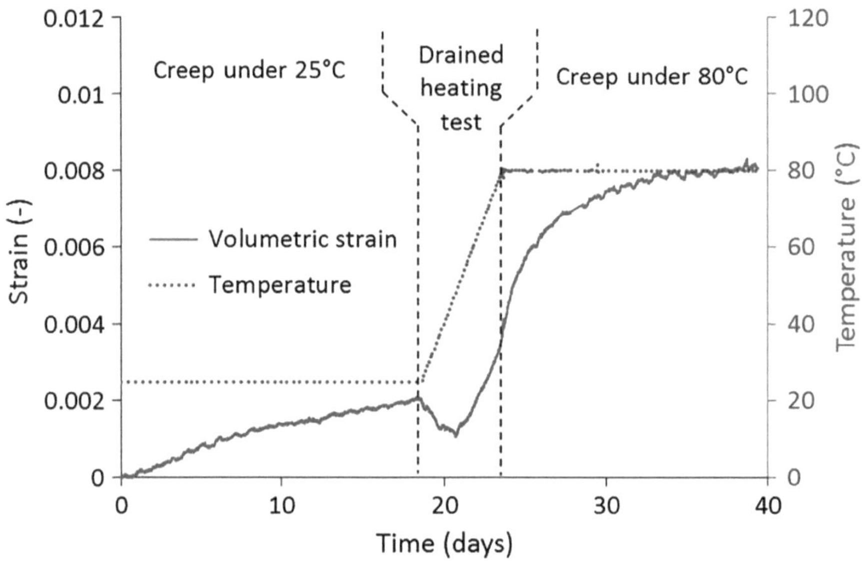

Fig. 1.11 Thermal effect on the creep behavior of COx claystone (Belmokhtar et al. 2017)

revealed that temperature increases can accelerate the creep deformation rate of clay rock. Cui et al. (2009) found a close relationship between the creep properties of Boom Clay and the overconsolidation ratio (OCR) value, based on creep tests conducted at different temperatures and stress paths. They observed that significant creep strain only occurred when the stress exceeded the preconsolidation stress, and the temperature was elevated. Towhata et al. (1993b) found that the creep deformation is initially accelerated during heating, meaning a faster decrease in void ratio or specific volume, but it is followed by a similar rate of deformation to that observed at lower temperatures. Burghingoli et al. (1992) found a positive linear relationship between changes in void ratio caused by thermal cycling and changes in void ratio caused by creep, indicating that increasing temperature may accelerate mechanical creep phenomena. In addition, thermal cycling will not cause an increase in void ratio when the change in void ratio caused by creep is zero.

Furthermore, other scholars (Fabre and Pellet 2006; Yu et al. 2015, 2022b, 2023b, c) have suggested the existence of a stress threshold for creep in clay rock. Djéran et al. (1994) noted that increasing temperature can lower the creep threshold of Boom Clay. However, Zhang and Rothfuchs (2004) and Lebon and Ghoreychi (2000), while testing COx claystone, arrived at different conclusions, indicating the absence of a significant creep threshold for this rock.

The creep behavior of clay rock is related to the thermo-mechanical path. As early as 1968, Campanella and Mitchell (1968) performed two groups of thermo-mechanical coupled drained triaxial creep tests on reconstituted illite and San Francisco Bay mud, i.e., initially consolidated at ambient temperature and the creep temperature. Their results showed that high temperature may accelerate the creep rate if the samples were initially consolidated at ambient temperature. Conversely, if the samples were initially consolidated at their creep temperature, the temperature has little effect on the time-dependent behavior. The results show that temperature effects on clay are dependent on the thermo-mechanical loading paths. This observation is further verified by Fox and Edil (1996), Zhang et al. (2007), and Cui et al. (2009). Fox and Edil (1996) found that the void ratio rate during creep increased exponentially with temperature, while cooling caused a decrease in creep deformation. However, Zhang et al. (2007) observed that heating induced the creep rate increase in COx claystone at the temperature range of 22–60 °C, while no creep deformation occurred during the cooling stage. In the reheating stage, the creep behavior was activated again when the temperature was increased to the highest temperature in the previous creep stage. Cui et al. (2009) found that thermally induced permanent deformation occurred when heating and cooling cycles were applied to normally consolidated or lightly overconsolidated clays, and it accumulated as the number of cycles increased. These observations indicate the link between thermally induced deformation and creep deformation.

1.2 Hydraulic Properties of Clay Rocks

1.2.1 Hydraulic Properties of Intact Clay Rocks

The transport properties of the host medium are significant factors for the safety assessment of the HLW disposal. Hydraulic conductivity and, hence, intrinsic permeability are parameters that can help verify the ease with which a host medium transmits fluids. As a sedimentary rock, the layered structures, like bedding in clay rock, can result in anisotropic hydraulic effects. To illustrate the anisotropic hydraulic properties, the hydraulic conductivity perpendicular (K_v) and parallel (K_h) to beddings, or intrinsic permeability k_v and k_h, can be identified through testing. Over the last few decades, several experiments, both in situ and in the lab, have been carried out to characterize the anisotropic hydraulic properties of the clay rocks, especially for the geological media for potential HLW geological disposal.

For instance, since the 1980s, when the High-Activity Disposal Experimental Site (HADES) was built in Mol, Belgium, to evaluate the feasibility of an HLW repository in Boom Clay formation, many researchers performed such experiments on Boom Clay (Bernier et al. 2007; Yu et al. 2013, Fig. 1.12). Diverse testing techniques, such as constant-head permeameter, tracer percolation, and isostatic investigations, have been employed to measure the hydraulic conductivities of Boom Clay. Most experimental studies exhibited similar Boom Clay hydraulic conductivities at the Mol site, on the order of 10^{-12} m/s, and the geometric means of K_h are 4.4×10^{-12} m/s, K_v is 1.7×10^{-12} m/s, with the anisotropy ratio K_h/K_v of roughly 2.5 (Yu et al. 2013). Similar experiments were conducted on Opalinus clay (Shao et al. 2011; Crisci et al. 2019; Al Reda et al. 2020; Hu et al. 2021), COx claystone (Distinguin and Lavanchy 2007; Harrington et al. 2013; Duan et al. 2021), and Tamusu mudstone (Yu et al. 2021b, 2023a). The hydraulic anisotropy K_h/K_v for the Opalinus clay is up to eight times, to about one order of magnitude (Shao et al. 2011). Both short-term and long-term measurements yielded hydraulic conductivity for COx claystone at different scales with considerable consistency, resulting in values less than 2×10^{-12} m/s (Distinguin and Lavanchy 2007). However, a discussion of the anisotropic ratio of COx claystone has not been found in the literature. The hydraulic conductivity for Tamusu mudstone is in the range of 10^{-13}–10^{-11} m/s, and its K_h/K_v deduced from the experimental results of Yu et al. (2023a) is about or even larger than one order of magnitude preliminarily.

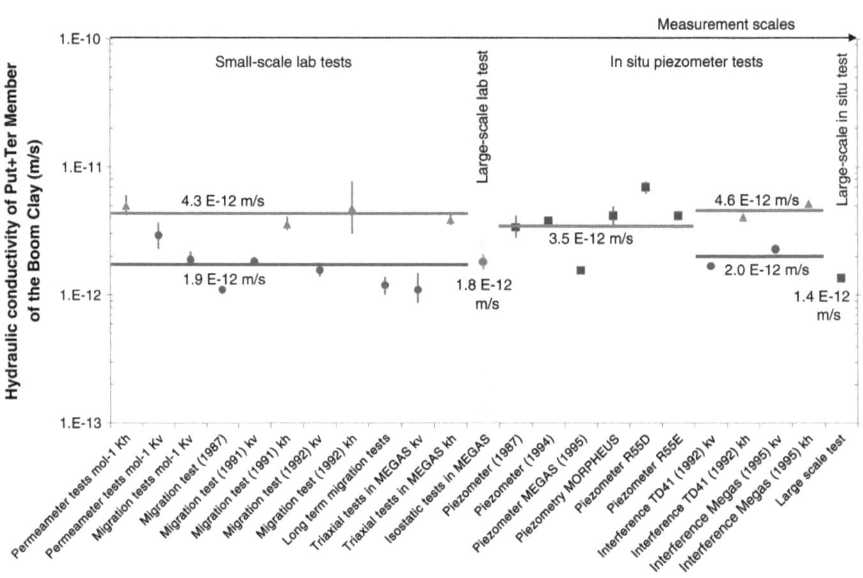

Fig. 1.12 Overview of hydraulic conductivity values (m/s) for Putte and Terhagen Members at the Mol site. Vertical bars represent the 95% confidence interval (Yu et al. 2013)

The rock near the HLW, exposed to elevated temperatures, suffers changes in its hydraulic properties. The relationship between hydraulic properties and temperature has been studied on many rock types. The results have varied, indicating direct (You and Kang 2009; Joshaghani and Ghasemi-Fare 2021), inverse (Yang et al. 2009; Liu et al. 2011; Brett-Adams et al. 2021), or even irrelevant relationships (Harrington et al. 2014; Zihms and Harrington 2015). Interestingly, Ren et al. (2014) pointed out that the silty clay's intrinsic permeability could decrease or increase depending on its dry bulk density in a thermal range of 10–25 °C. Guo et al. (2017) performed experiments on three kinds of rocks, which were specifically identified as low-, medium-, and ultra-low permeabilities. They found that permeability possessed a negative nonlinear correlation with temperature for low and medium-low permeability rocks and a positive nonlinear correlation for ultra-low permeability rocks. Although considerable research has been performed on the temperature effect on the hydraulic characteristics of geomaterials, results related to clay rocks, especially those under-considered for nuclear waste underground disposal, are limited. Several experimental studies on Boom Clay have indicated that temperature could significantly affect its hydraulic properties (Romero et al. 2001; Chen et al. 2014, Fig. 1.13; Ma et al. 2016). As reported by Sultan (1997), Boom Clay had a twofold hydraulic conductivity at 35–60 °C . This thermal enhancement phenomenon in the permeability of Boom Clay is further verified by Delage (2009), Chen et al. (2014),

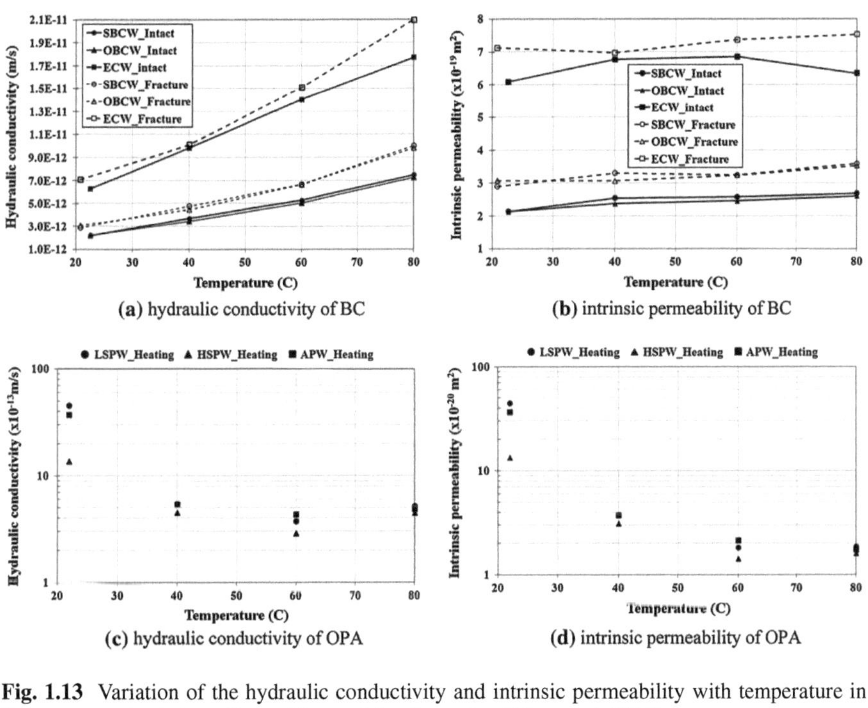

Fig. 1.13 Variation of the hydraulic conductivity and intrinsic permeability with temperature in the permeameter tests on Boom Clay and Opalinus clay (Chen et al. 2014)

and Chen et al. (2017). Similar results were obtained from studies on Opalinus clay (Monfared et al. 2011a, 2014).

1.2.2 Hydraulic Properties of Fractured Clay Rocks

In situ observations have shown that repository excavation induces damage in the host rock in the form of fracture networks (Coll et al. 2004; Mertens et al. 2004; Bastiaens 2005, Bastiaens et al. 2007; Bernier et al. 2007; Armand et al. 2014). The zone containing these artificial fracture networks is known as the excavation damage zone. The flow and solute transport in clay rocks are mainly controlled by natural and artificial fracture networks. These fractures can be affected by many factors, such as the temperature rise caused by the decay of radioactive nuclides and stress changes during the construction and operation of the disposal repository. Understanding the hydraulic changes of fractured clay rocks during the THM process is crucial for understanding the flow and transport of fluids within fractures, as well as further analyzing the migration patterns of nuclides and evaluating the safety of disposal repositories.

The studies from Fujii et al. (2015) show that in fractures, permeability was elevated by 1–3 orders. Then, shear fracture zone morphologies were detected by X-ray CT, which suggested that the alterations of permeability detected were due to smoothing of the shear zone, fracture generation, reactivation, and closure at diverse stresses. Houben et al. (2020) suggested that for Whitby mudstone, its increased permeability was under the control of fracture permeability, and it showed high sensitivity to the hydro-chemo-mechanical responses of fracture to fluids. But smaller cracks were detected using scanning electron microscopy (SEM) and X-ray microtomography, which were associated with the percolating network. The above-mentioned research shows the mechanical loading effect on the permeability variation of fractured mudrocks from both macro and micro perspectives. Duan et al. (2021) presented that the permeability evolution of COx claystone due to hydric cycles is attributed to the occurrence. The permeability of the "damaged" zone is two to three orders of magnitude more than what is likely to be representative of the matrix permeability (i.e., non-cracked material) (Fig. 1.14).

In addition to the mechanical and thermal effects, another key issue that is closely related to the transport characteristics of the host rock is the self-sealing effect of fractured clay rock. Self-sealing, associated with the reduction of fracture transmissivity, is widely observed in clay rocks, especially in the HLW disposal host rocks (Figs. 1.15, 1.16, 1.17 and 1.18), such as Boom Clay, COx claystone, and Opalinus clay, etc. The experimental studies performed by Chen et al. (2014) on Boom Clay and Opalinus Clay prove that confining pressure may contribute to their fracture self-sealing and the reduction of their permeability. Using microfocus X-ray CT, they observed that Opalinus clay showed a significantly weakened sealing compared with Boom Clay. Similarly, Van Geet et al. (2008) carried out permeability tests on Boom Clay and Opalinus clay with planar fractures of about 300 μm

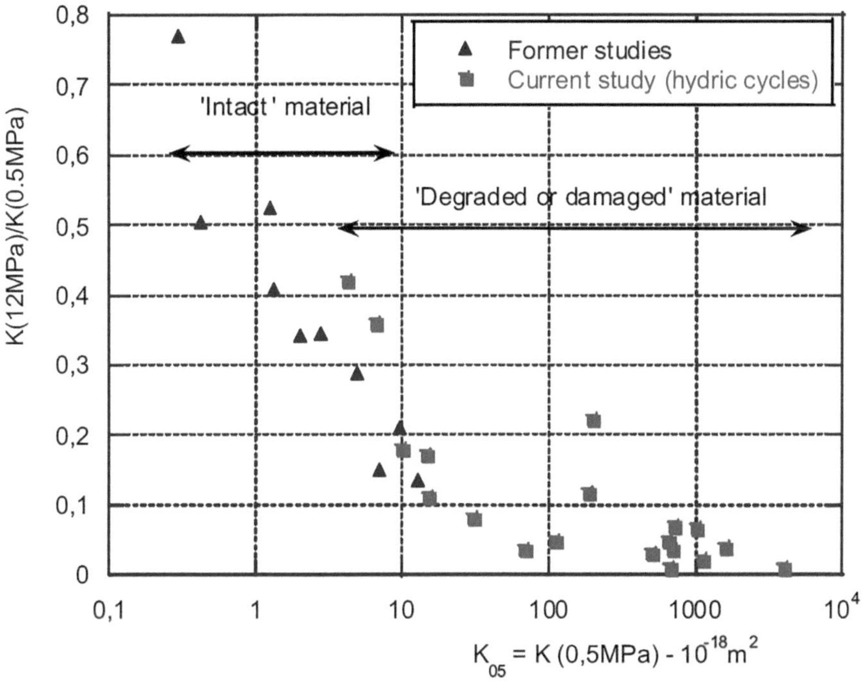

Fig. 1.14 Sensitivity of gas permeability to confining pressure vs. permeability at 0.5 MPa. (Duan et al. 2021)

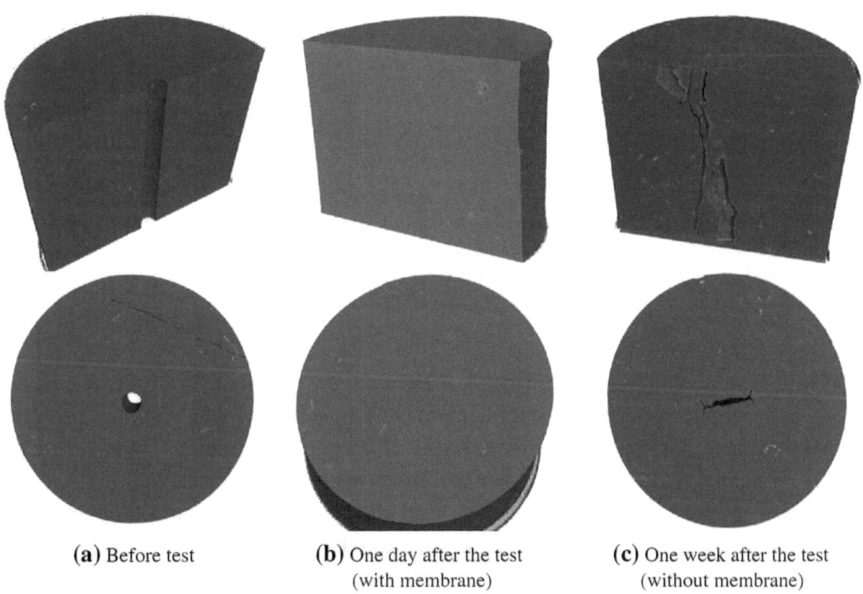

Fig. 1.15 μCT for the Boom Clay sample before (a) and after (b) the isostatic test, and after dismantling and exposing to the air for a couple of days (c) (Chen et al. 2014)

Fig. 1.16 μCT scans for the Opalinus clay sample injected with LSPW before and after the permeameter test (Chen et al. 2014)

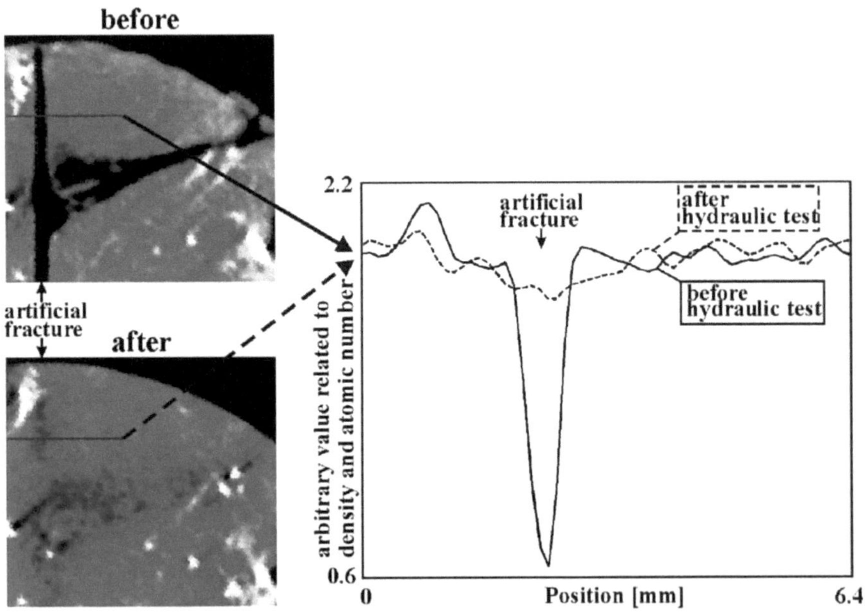

Fig. 1.17 Evolution of the fracture aperture at the top of a Boom Clay sample tested with SBCW (Synthetic Boom Clay Water) for permeability evolution. Notice the disappearance of the aperture, but the small decrease in density over a zone of twice the original fracture aperture. (Van Geet et al. 2008)

aperture. Using a microfocus X-ray CT technique, the authors discovered that Boom Clay had fully recovered hydraulic characteristics (Fig. 1.17). On the contrary, Opalinus clay just had mildly recovered characteristics. Zhang and Rothfuchs (2008), Zhang (2011, 2013) observes that fracture closure and permeability reduced to 10^{-19}–10^{-21} m^2 in a few months to several years for both Opalinus clay and COx claystone in the THM environment. As suggested by Monfared et al. (2012), permeability was not markedly affected by the shear band, thus verifying the favorable self-sealing characteristics of Boom Clay under high and room temperatures. Zhang

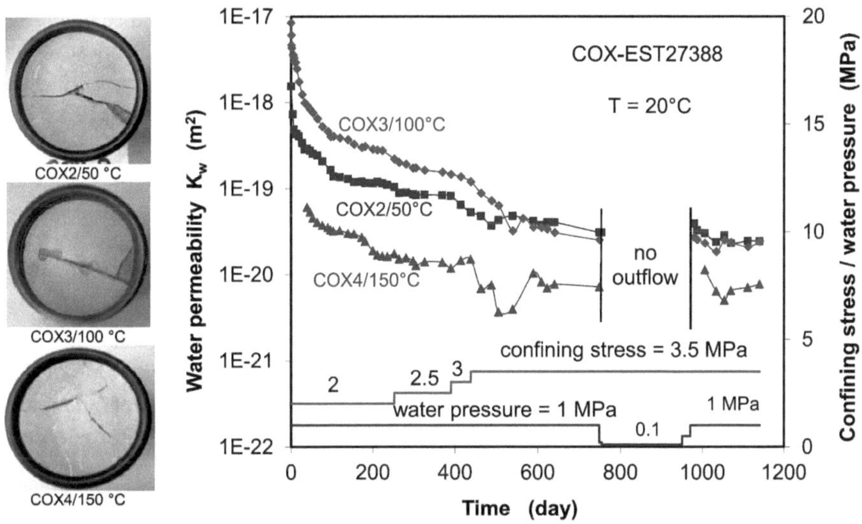

Fig. 1.18 Long-term measurements of water permeability on fractured COx claystone samples preheated at 50 °C, 100 °C, and 150 °C (Zhang et al. 2017)

et al. (2017) reported that the preheating up to 150 °C did not hinder the sealing of the fractures in the claystone. The final permeability values determined after 3 years are the same as those of the intact clay rock (Fig. 1.18). In situ experiments and observations so far at different Underground Research Laboratories (URLs), such as the Meuse/Haute-Marne URL in France (COx claystone), HADES URL in Belgium (Boom Clay), and Mont Terri URL in Switzerland (Opalinus clay), exhibit the distinct self-sealing process. "Softer" clay rocks, including Boom Clay, show a rapid reaction and display fracture closure (sealing) and even healing signs. As for "hard" clay rocks (COx claystone, Opalinus clay), it needs a significantly longer time for self-sealing (Bühler et al. 2003; Bossart et al. 2004; Blüemling et al. 2007; Bastiaens et al. 2007; Van Marcke and Bastiaens 2010; De la Vaissière et al. 2015; Voltolini and Ajo-Franklin 2020).

1.3 Mechanisms of Thermal Effects on Hydro-Mechanical Properties of Clay Rocks

Many efforts have been proposed to explain the mechanism of the thermal effects on the clay rocks' properties. Researchers explained the thermal effects in terms of changes in clay fiber characteristics, pore water viscosity, and particle–water interaction. However, opinions are divided on this topic, and no consensus has been reached.

The driving mechanisms behind the thermal response of clay rock could be attributed to various physical phenomena occurring in the clay microstructure. For instance, Pusch and Güven (1990) investigated the microscopic structure change of a hydrothermally treated bentonite clay using scanning electron microscopes (SEM). They observed that the flake stacks of clay disintegrated during heating and yielded dense branches of stacks separated by large voids. Pons et al. (1994) and Thomas et al. (1994) found that high temperature can result in a structural reorganization of montmorillonite gel. Almanza et al. (1995) observed that the elevated temperature promoted the rearrangement of clay stacks and caused flocculation in clay. However, the temperature effect on the microstructure of clay rock has not been fully understood because of the limitations of conventional experimental techniques. Romero et al. (2001) suggested that the increasing temperature leads to clay fabric alteration and porosity redistribution by thermo-chemical effects. Chen et al. (2017) observed that heating induced changes in clay fabric and resulted in porosity redistribution in clay rock, with larger voids detected through nuclear magnetic resonance tests. Houhou et al. (2021) presented a first attempt at microstructure observations in compacted clays subjected to thermal loading. They found that the macropores in the heated illite sample are smaller and more dispersed than those in the unheated one (Fig. 1.19). The thermal contraction of illite could be attributed to the deformation/collapse of macropores in the dual-porosity structure assemblage.

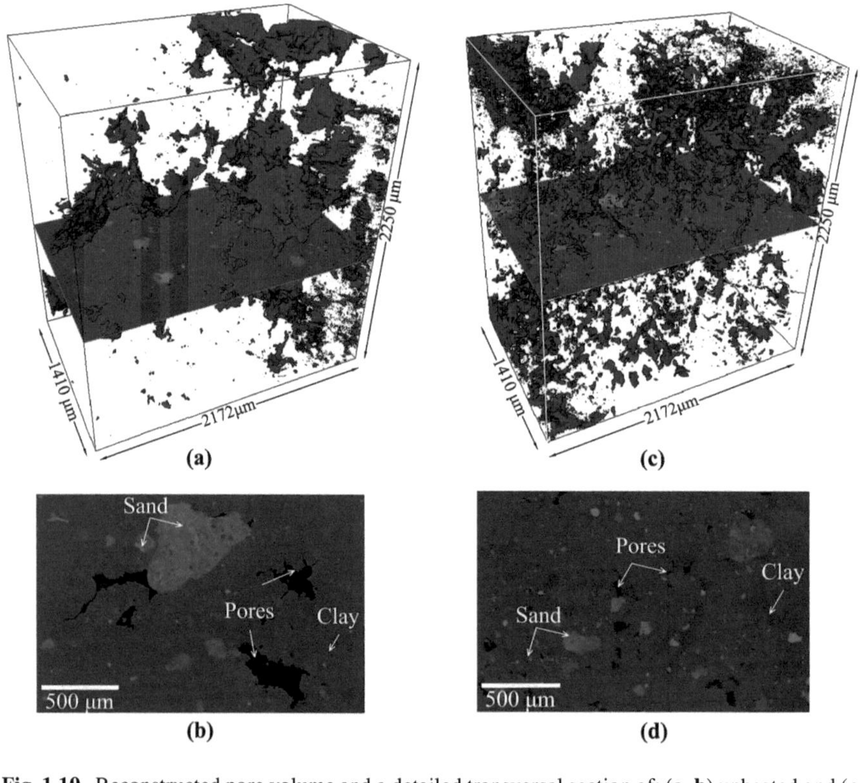

Fig. 1.19 Reconstructed pore volume and a detailed transversal section of: (**a**, **b**) unheated and (**c**, **d**) heated Illite samples (Houhou et al. 2021)

Furthermore, the expansion of minerals can be anisotropic along their crystallographic axes (c-axes). Tribaudino et al. (2010) demonstrated that plagioclase feldspars expanded 70% more along their c-axes compared to their other axes. This anisotropic expansion of minerals can contribute to understanding the deformation mechanism of clay rock during heating. More work needs to be done on this issue due to its significant impact on clay thermo-hydro-mechanical properties.

Water in saturated clays is presented as free water and clay-bound water (Fig. 1.20). The difference is due to the electrochemical interaction between water and clay particles. As shown in Fig. 1.20, free water is the amount of water that fills the clay pores and is known as bulk water in the soil skeleton, while bound water is absorbed by the surface of clayey minerals and plays a significant role in the thermo-hydro-mechanical behavior of clayey materials. Some scholars consider that the changes in electrical double layer thickness could be closely related to the thermal effects. Since the size of the clay particles is very small and particle surfaces are usually negatively charged, clays are characterized by strong colloid properties when exposed to water. Polarized water molecules and ions are strongly attached to the surfaces of clay particles via the Coulomb force and form the diffusive double layers. The schematic diagram of the double layer in the vicinity of clay particles is shown in Fig. 1.21. The thickness of double layers may affect the Coulomb force (repulsion) between clay particles, thus changing the mechanical properties of clays.

When clays are heated, the bonding between the clayey minerals' surface and water is weakened, and some of the bound water transforms into free water under certain temperature and pressure conditions (Baldi et al. 1988; Sojoudi and Li 2023). Campanella and Mitchell (1968) interpreted the thermal softening effect of clays in terms of the thermal motion of molecules. They noted that the elevated temperature indicates that temperature would increase the kinetic energy of the

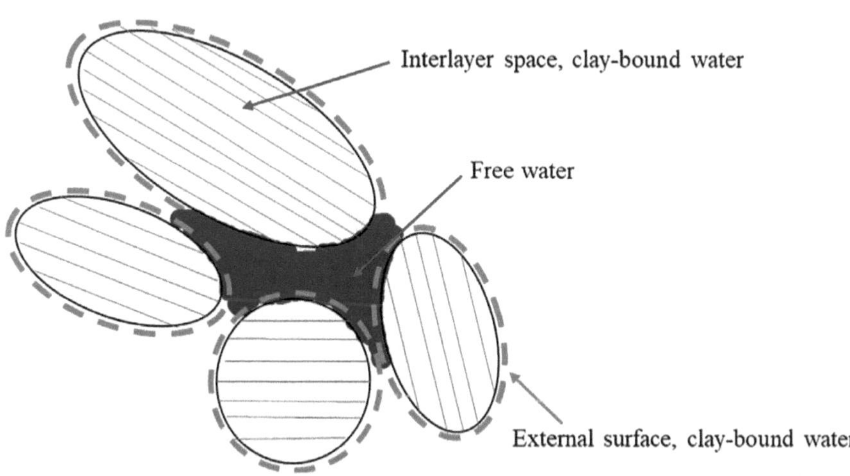

Fig. 1.20 Sketch showing free water and bound water in clay-water composites (Sojoudi and Li 2023)

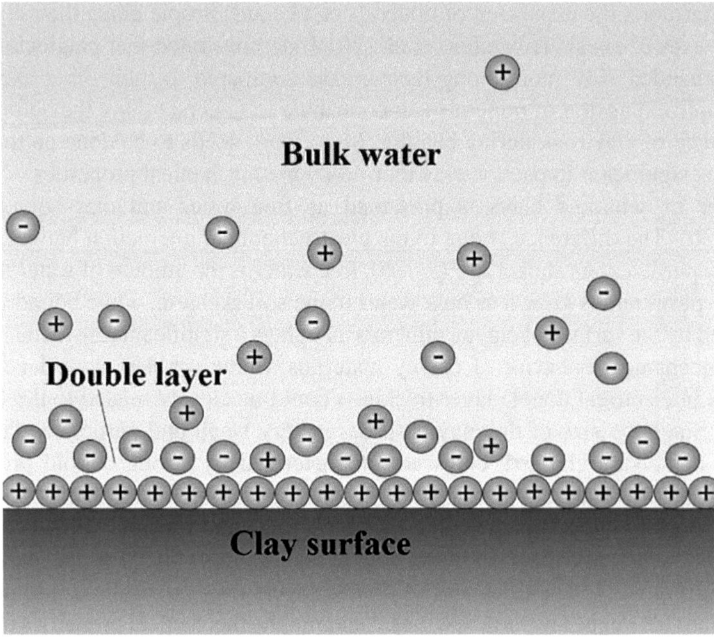

Fig. 1.21 Double layer in the vicinity of clay particles (Wikipedia 2015)

molecules and weaken the bonds between the clay particles, thus resulting in a reduction of clay strength. Baldi et al. (1988) interpreted the thermal volume change behavior of low-porosity clay from the perspective of the thermal expansion coefficient difference of adsorbed water and free water. Robinet et al. (1996) investigated the effect of temperature on the structure of the clay from the three classes of pore spaces, i.e., the interlamellar space, the interparticle space, and the interaggregate space. The consolidation of clay is mainly caused by the reduction of the interaggregate space at room temperature. While at elevated temperature, the thermal agitation of adsorbed water would leave an unstable contact of clay particles, giving rise to the compaction of the interlamellar space.

However, thermal effects on the thickness of the double layer are still controversial. Weaver (1979) suggested that elevated temperatures would result in an acceleration of ion movement and improve the solubility of the electrolyte, thus resulting in an increase in ion concentration in the vicinity of the clay particles and a reduction of the double layer. Morin and Silva (1984) considered that increased temperature would affect the pH value of pore water and decrease the thickness of the electric double layer. These hypotheses were confirmed by Tidfors and Sallfors (1989) and Wang et al. (2010) through experiments. However, contradictory conclusions have been presented in the literature. Yong (1975), Plum and Esrig (1969), and Derjaguin et al. (1986) observed that increasing temperature decreased the double layer thickness. Mitchel (1976) considered that temperature reduced the dielectric constant of water, thus balancing the expansion trend of the double layer. Later, Towhata et al. (1993a) explained that elevated temperature would prompt the water molecules to escape from the range of the Coulomb force. Therefore, a reduction of adsorbed water would occur, giving rise to the closer contact of clay particles. Hence, no overall change would occur in the thickness of the double layer. Jefferson (1994) interpreted this contradictory observation. He considered that all these studies are based on the Gouy-Chapman double layer theory. It is assumed in the theory that temperature is an independent variable. However, the double layer thickness is dependent on many other variables, such as pH value, ion concentration, and the polarity characteristic of water molecules. All these variables are intercoupling.

The drained creep of clay rock is related to two processes: dissipation of pore water pressure and viscosity of the solid skeleton. Heating may produce an increase in the hydraulic conductivity of saturated clay, thus accelerating the dissipating rate of pore water pressure. To distinguish whether the creep behavior is due to pore pressure dissipation or the viscosity of the solid skeleton, Djéran et al. (1994) conducted a set of undrained creep tests at different temperatures. The results clearly showed that elevated temperature increased the viscosity of the clay skeleton and decreased the creep threshold.

The thermal effect on clay rock is a complex phenomenon that involves many factors, including but not limited to the thermal expansion of the soil–water system, the hydration of soil colloids, the arrangement of clay particles, the thermal motion of molecules, and the thermo-mechanical history. These factors may counteract each other, which makes it difficult to distinguish the dominant mechanisms. The existing studies cannot fully explain the mechanism behind the observed THM behavior of clay rocks. More in-depth research on thermal effects is required.

1.4 Theoretical Studies on the THM Behavior of Clay Rocks

1.4.1 Thermo-Mechanical Constitutive Models for Saturated Clay Rocks

Some researchers have developed thermal-related constitutive models that can reproduce most of the observed behavior of saturated clay rocks at high temperatures. The first conceptual model for thermal effects on the strain response of clay was given by Campanella and Mitchell (1968). Schiffmann (1971), Derski and Kowalski (1979), and Bear and Corapcioglu (1981) conducted research on the thermal consolidation of clay rock. However, these early studies did not consider the irreversible strains explicitly.

In 1990, Hueckel and Borsetto discussed the plastic behavior of clays due to heating and loading under constant elevated temperature and developed thermoplasticity constitutive models for clays by extending the well-known modified Cam-Clay (MCC) model to consider thermo-elasto-plastic behavior (Fig. 1.22). The yield

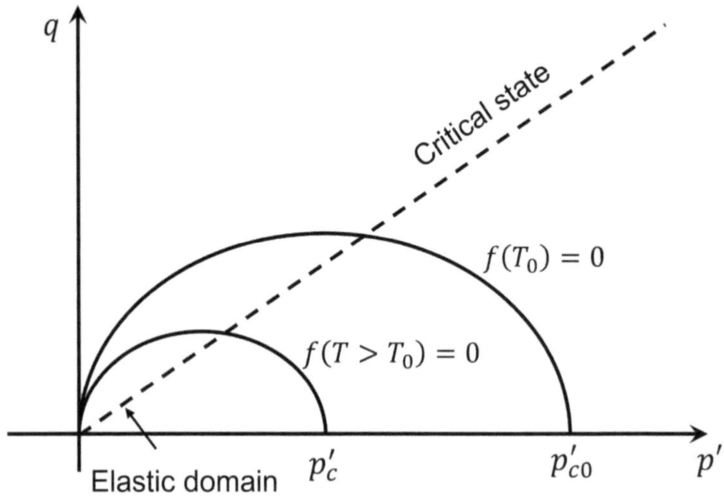

Fig. 1.22 Yield surface shrinking due to temperature at elastic state (Hueckel and Borsetto 1990)

surface is assumed to be a function not only of stress and plastic volumetric strain, but also of temperature difference ΔT (referred to as an environment temperature T_0) (1.1),

$$f = f\left(p', q, \varepsilon_V^{tp}, \Delta T\right) \quad \Delta T = T - T_0, \tag{1.1}$$

where the effective stress invariants p' and q are defined for the six-component effective stress, and ε_V^{tp} is the thermo-plastic volumetric strain. The proposed model accounts for the shrinking of the elastic domain during heating (thermal softening) and expansion during cooling when the stress state is within the yield surface.

In their companion paper, Hueckel and Baldi (1990) extended the well-known Modified Cam-Clay (MCC) model by introducing a yield surface considering temperature effects on the preconsolidation pressure to reproduce the thermal-induced plastic strain (Fig. 1.23):

$$p'_c = p'_c\left(\Delta T, \varepsilon_V^p\right), \tag{1.2}$$

where p'_c is the preconsolidation isotropic pressure, ε_v^p is the volumetric plastic strain, $\Delta T = T - T_0$, T is the given temperature, T_0 is the reference temperature.

Later, the thermoplastic theory was combined with a model for thermoelastic expansion of interstitial water in low-porosity clays (Hueckel and Pellegrini 1991), and it was further improved in 1998 to include the variability of the carbonate content by Hueckel et al. (1998).

Adopting similar assumptions, subsequent models accounting for thermal effects are based on the same approach, utilizing experimental data from different researchers within the framework of the Cam-Clay model. Robinet et al. (1996) proposed a

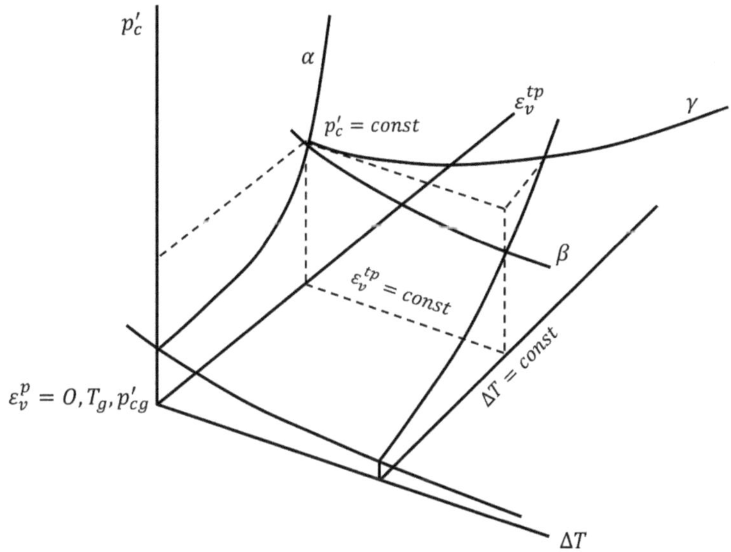

Fig. 1.23 Apparent preconsolidation isotropic pressure p'_c as a function of ε_V^{tp} and ΔT (Hueckel and Baldi 1990)

thermo-mechanical constitutive model for non-expansive saturated clays, characterizing the mechanical yield surface based on the MCC model and the thermal softening yield surface as a plane separating two thermal domains. The preconsolidation pressure of clay is a function of mechanical plastic volumetric strain and thermal expansion volumetric strain:

$$p_c = p_{c0}\exp\left(-\beta_m \varepsilon_v^p\right)\exp\left(\beta_T \varepsilon_{Ts}^p\right) \tag{1.3}$$

where p_{c0} is the initial value of preconsolidation pressure, ε_{Ts}^p is the irreversible expansion thermal strain, and β_m, β_T are the model parameters.

Cui et al. (2000, 2006) proposed a thermo-mechanical model for saturated clays subjected to temperature changes, but with particular attention to the volume change behavior and the effects of the overconsolidation ratio. Later, Abuel-Naga et al. (2007) developed a new TY surface based on the multi-yield surfaces model of Cui et al. (2000, 2006). This model was based on the test results on Bangkok soft clay and has fewer parameters than the model of Cui et al. (2000). However, the temperature has no effect on preconsolidation pressure in this model, which contradicts experimental data of most clays. Graham et al. (2001) proposed a thermo-mechanical MCC model by defining the compression index, swelling index, and preconsolidation pressure as functions of temperature. Laloui and François (2009) developed a multi-yield-surface thermoplastic model and introduced a unique yield surface to reproduce the thermo-mechanical behavior at high and intermediate OCRs (overconsolidation ratio) (Fig. 1.24). This model offers a more advanced concept

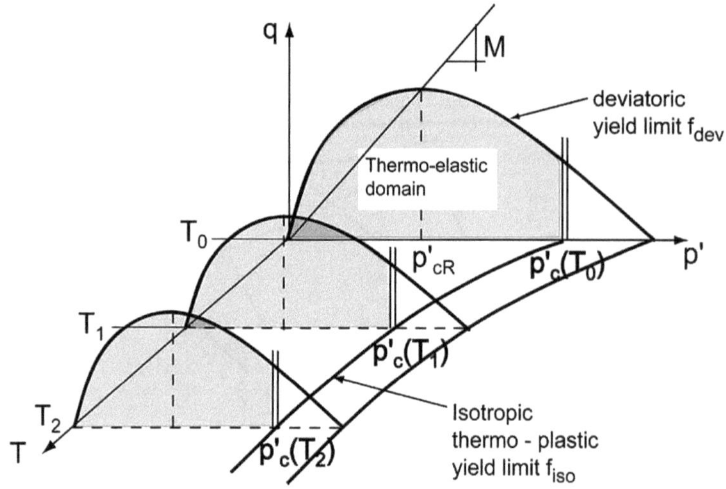

Fig. 1.24 Coupled thermoplastic yield limits proposed by Laloui and François (2009)

combining the efficiency of the isotropic thermoplastic mechanism and the capability of a Cam-Clay family deviatoric mechanism enhanced by the bounding surface theory. Abuel-Naga et al. (2009) proposed a thermoelastic-plastic (TEP) mechanical model for saturated clays in triaxial stress space at states wetter than critical under the framework of critical state soil mechanics and the MCC model. Yao et al. (2011a, b) extended the Cam-Clay model and unified hardening model for non-isothermal clay by considering the temperature-dependent value of the friction angle (or slope of critical state line) and preconsolidation pressure (Fig. 1.25).

Zhang et al. (2012) introduced temperature-deduced equivalent stress into the subloading yield surface model. This new concept is used to formulate an equivalent subloading yield surface, an extended void ratio difference, and the corresponding evolution equation in a natural way, which can consider the influences of both the stress and temperature. However, the temperature effect on preconsolidation pressure was ignored in this model. Hamidi et al. (2015) first developed a thermo-mechanical constitutive model for saturated clays based on the general framework of critical state soil mechanics and modified Cam-Clay formulation. In their model, an attempt has been made to use the lowest possible number of extra parameters compared with the original Cam-Clay model and to ensure that these new parameters have clear physical interpretations. Two years later, they extended the constitutive model by considering the effects of structure on the mechanical behaviors of the clay (Hamidi et al. 2017). Sojoudi and Li (2023) proposed a thermodynamic-based model for modeling thermo-elasto-plastic behaviors of saturated clays considering bound water dehydration. In their models, the dehydration strain is described using a novel method based on a generalized thermodynamics approach and the Helmholtz free energy function. The irreversible strain is captured using Cam-Clay plasticity and with the subloading yield surface concept.

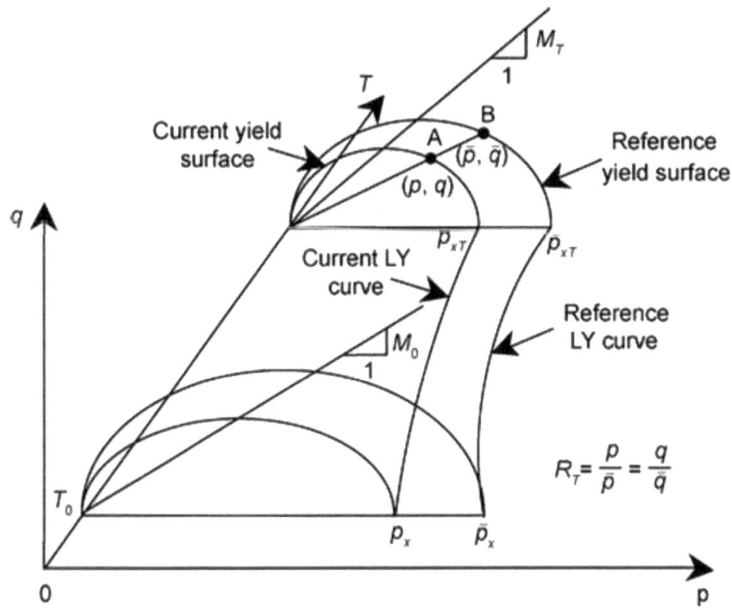

Fig. 1.25 The yield surface developed by Yao et al. (2011a) and the reference yield surface

Many of the thermo-mechanical models, as mentioned above, are formulated within the framework of critical state soil mechanics, assuming the yield function of the MCC model, in which thermoplastic behavior is incorporated using a temperature-dependent yield pressure. However, for indurated clay rocks that have undergone strong post-sedimentary diagenesis, these models may not necessarily be applicable. Therefore, from a practical standpoint, other failure criteria, such as Drucker–Prager and Mohr–Coulomb types, may be applicable for characterizing the behavior of the stiff clayey materials considered as possible host mediums for deep geological disposal (Baldi et al. 1991; Labiouse 1997; François et al. 2011; Yu et al. 2014). Laloui and Cekerevac (2003) present a thermoplastic mechanism for isotropic thermo-mechanical paths of clay rock. In the model, an isotropic yield mechanism including thermal hardening is proposed based on considerations of the thermal effect on the void ratio. Liu and Xing (2009) proposed a thermo-elasto-plastic constitutive model based on a double hardening model for clays, which describes the combined effects of temperature and OCR on the mechanical properties of saturated clays. Gong et al. (2015, 2016) developed a thermo-elasto-plastic damage model based on the Drucker–Prager Cap model. The model involves several temperature-dependent parameters, i.e., the cohesion, the elastic modulus, and the preconsolidation pressure, to capture the mechanical behavior of Boom Clay at various temperatures. Based on a Mohr–Coulomb type failure criterion, Mánica et al. (2017) present a thermomechanical constitutive model for clay rocks with the consideration of their anisotropy of strength and stiffness, behavior nonlinearity, occurrence of plastic strains before peak strength, significant softening after peak, time-dependent creep deformations, and permeability increase due to damage. Zhang and Cheng (2016, 2017) developed the Tsinghua-Thermosoil (TTS) model based on a non-equilibrium thermodynamic approach (Fig. 1.26). In the model, there are no concepts of yield surface and flow rules in the constitutive relations. Instead, it uses the Onsager migration coefficient model, determines the elastic potential energy function, and introduces the concepts of granular entropy and granular temperature with the consideration of granular fluctuation. Song et al. (2025) developed a fully coupled THM constitutive model for clay rocks based on the Mohr–Coulomb criterion and Perzyna's overstress theory (Fig. 1.27). The constitutive model considers nonlinearity and softening after peak strength, anisotropy of stiffness and strength, as well as permeability variation due to damage.

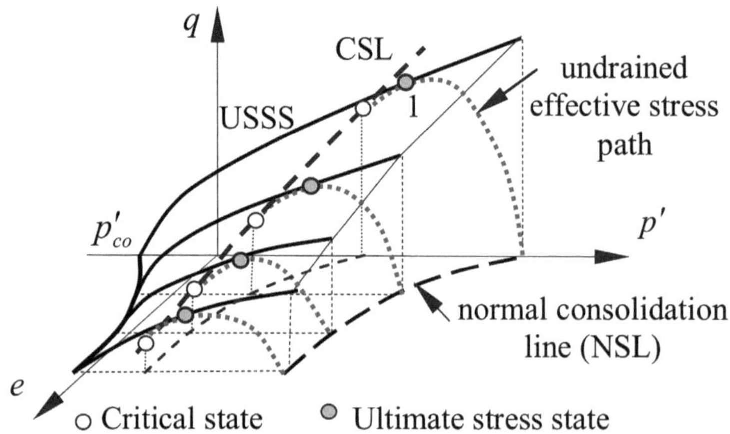

Fig. 1.26 Schematic diagram of the ultimate stress state surface and the critical state in e–p'–q space of Tsinghua-Thermosoil model (Zhang and Cheng 2017)

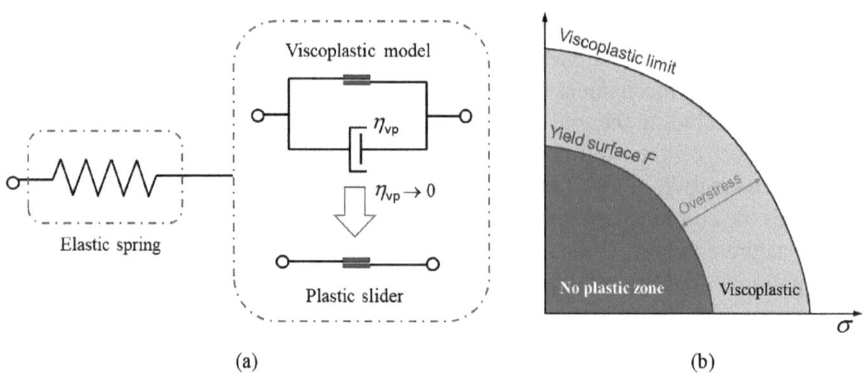

Fig. 1.27 Conceptual model of (a) the adopted elasto-viscoplastic constitutive model and (b) the overstress formulation of the viscoplastic model (Song et al. 2025).

The existing thermoelastic constitutive models have effectively described some experimental and on-site monitoring results. However, many thermal elasto-plastic constitutive models are macroscopic and typically consider temperature based on extensions of the Cam-Clay model. Their effectiveness beyond experimental results is questionable due to the lack of consideration for microscale physical processes. In addition, most constitutive models have certain limitations in predicting the thermal response mechanism of clay through long-term heating and cooling cycles.

1.4.2 Creep Models for Clay Rocks Considering Temperature Effects

Valuable research results have been achieved in the theoretical study of the creep behavior of clayey rocks. However, constitutive models that are able to consider the temperature effect on the creep behavior or the time effect on the thermomechanical behavior of clays are still limited. As early as 1972, Meschyan and Galstyan proposed a creep model that directly introduced the temperature as a parameter in the expression of creep rate. To some extent, this simple approach can reproduce the thermal effects on the creep. However, it cannot tackle the mechanism of these effects. The model is applicable only when the load and temperature are kept constant because the explicit time is included in the expression. This limitation significantly hindered its applications. Modaressi et al. (1997) developed a viscoplastic model based on the Perzyna model, wherein the thermal effects on friction angle, preconsolidation pressure, and viscosity are explicitly considered. This model can tackle the creep behavior of clay under different comprehensive thermo-mechanical paths. The creep rate can be expressed as:

$$\dot{\varepsilon}_{\mathrm{vp}} = \frac{1}{\mu}\left\langle \frac{F_{\mathrm{vp}}}{f_0'} \right\rangle^n \frac{\partial G_{\mathrm{vp}}}{\partial \sigma} \tag{1.4}$$

in which

$$\mu = \eta \exp\left(\frac{\zeta}{T}\right) \tag{1.5}$$

$$F_{\mathrm{vp}} = q - pH_{\mathrm{v}}H_{\mathrm{r}} \sin(\phi - kT) \tag{1.6}$$

where μ is the viscosity of clay, F_{vp} is the overstress, f_0' is the reference pressure, G_{vp} is the creep potential function, H_{v}, H_{r} are the hardening parameters, ϕ is the friction angle, and η, ζ, k are the model parameters.

Tang et al. (2008) combined the Perzyna model with the multi-yield-mechanism model of Cui et al. (2000, 2006) and formulated a new thermo-elasto-viscoplastic model to deal with the time-dependent behavior of clays. The strain was separated into elastic and viscoplastic components:

$$d\varepsilon_{\mathrm{vT}} = d\varepsilon_{\mathrm{vTe}} + d\varepsilon_{\mathrm{vTvp}} \tag{1.7}$$

in which

$$\mathrm{d}\varepsilon_{\mathrm{vTvp}} = \gamma_{\mathrm{T}} \left(\frac{T - T_{\mathrm{CT}}}{T_{\mathrm{CT}}} \right)^{\alpha_{\mathrm{T}}} \mathrm{d}T \tag{1.8}$$

where γ_{T}, α_{T} are the model parameters, and T_{CT} is the yield temperature.

Kurz et al. (2016) present a semi-empirical elastic-thermoviscoplastic model for clay, which combines thermoplasticity with viscoplasticity into an elastic-thermoviscoplastic (ETVP) model (Fig. 1.28). Strains were partitioned into elastic (recoverable) and plastic (nonrecoverable) components. The plastic component includes viscous strains defined by a creep rate coefficient that varies with plasticity index and temperature (T), but not with stress level or overconsolidation ratio (OCR).

Qiao and Ding (2017) proposed a thermoviscoplastic constitutive model to describe the viscoplastic behavior of geomaterials with the consideration of thermal

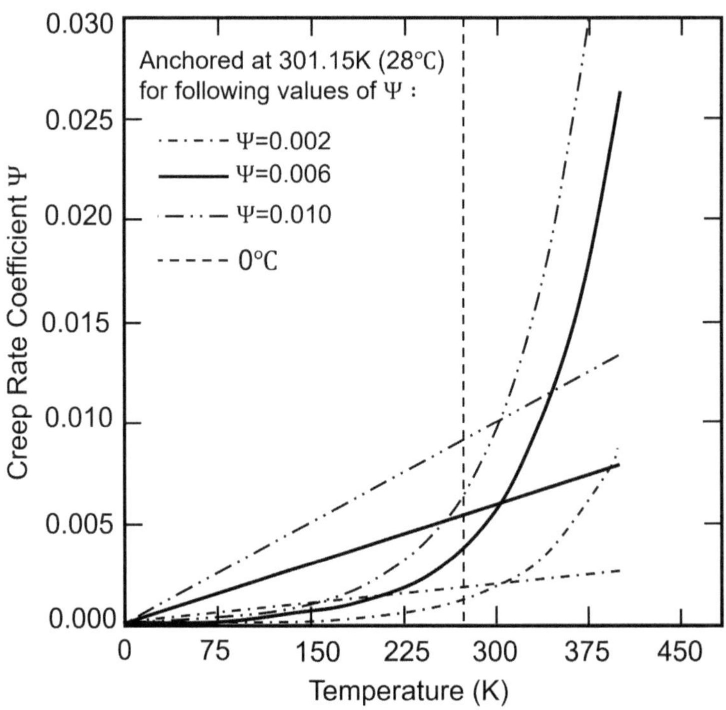

Fig. 1.28 Linear and exponential relationships for the creep rate coefficient ψ versus temperature anchored at a temperature of 28 °C (301.1 K) for given values of ψ (Kurz et al. 2016)

effect (Fig. 1.29). The viscous behavior, formulated in terms of strain rate, is described using the nonstationary flow surface theory. The yield function is written as follows:

$$f = \left(\sigma'_{ij}, \alpha_m, \varepsilon_{ij}^{vp}, \dot{\varepsilon}_{vol}^{vp}, T\right) \tag{1.9}$$

where σ'_{ij} is the effective stress, α_m is the vector of non-hardening parameters, ε_{ij}^{vp} is the viscoplastic strain, $\dot{\varepsilon}_{vol}^{vp}$ is the viscoplastic volumetric strain rate, and T is the temperature. The temperature influence is introduced by adopting a temperature-dependent mean effective preconsolidation pressure.

$$p'_c = p'_{c0} \exp\left(\beta \varepsilon_{vol}^{vp}\right) \tag{1.10}$$

$$p'_{c0} = p'_{c0,ref} \left(\frac{\dot{\varepsilon}_{vol}^{vp}}{\dot{\varepsilon}_{vol,ref}^{vp}}\right)^{C_A} \left[1 - \gamma \log\left(\frac{T}{T_{ref}}\right)\right] \tag{1.11}$$

where p'_{c0} is the mean effective preconsolidation pressure at strain rate $\dot{\varepsilon}_{vol}^{vp}$ and temperature T with a viscoplastic strain 0, and $p'_{c0,ref}$ is the mean effective preconsolidation pressure at the reference strain rate $\dot{\varepsilon}_{vol,ref}^{vp}$ and reference temperature T_{ref} with a viscoplastic strain of 0. β is the viscoplastic stiffness, which is the reciprocal of the slope of the critical state line in the $\left(\varepsilon_{vol}^{vp} - \ln p'\right)$ plane. C_A is a material parameter related to the viscosity.

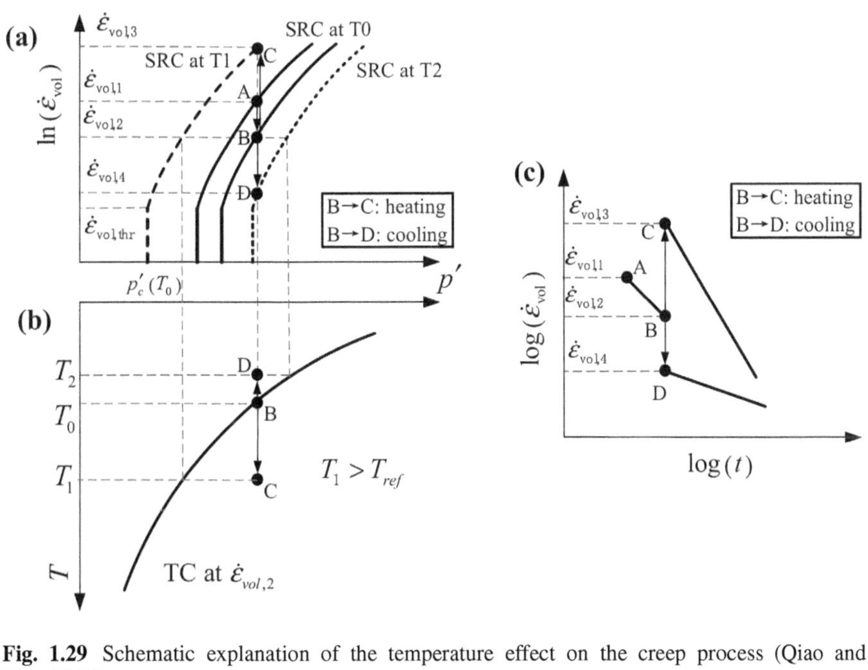

Fig. 1.29 Schematic explanation of the temperature effect on the creep process (Qiao and Ding 2017)

The TTS model was developed for clay, where thermal strains are related to the underlying physical mechanism of conversion of bound water to free water within the clay (Zhang and Cheng 2016, 2017). It provides a more realistic representation of the long-term response of clay due to continuous heating and cooling, using a small number of easily calibrated parameters (Zymnis et al. 2018, 2019; Zymnis and Whittle 2020). Based on this model, Zymnis et al. (2019) investigate the capabilities of the TTS model in simulating familiar aspects of thermal consolidation of clays, as well as the long-term, progressive accumulation of strains associated with seasonal heating and cooling processes for shallow geothermal systems installed in clays.

Shao et al. (2018) put forward the creep equation of the Burgers model in the variable temperature field, established the improved elastic and viscous elements, so that the improved Burgers model can be applied to the condition of temperature-stress seepage coupling of clay rock. Yang (2018) introduced the damage factor based on a traditional rheological model and described the temperature-stress coupled creep damage constitutive model of sandy clay rock.

1.4.3 Hydraulic Conductivity Evolution Model Considering Thermal Effects

Several theoretical studies have been conducted to characterize the evolution of hydraulic conductivity for geomaterials, taking into account factors, such as porosity, stress, temperature, and chemical processes. Generally speaking, there are four main categories of hydraulic conductivity evolution models, namely based on (i) porosity (Carman 1937; Li 2020; Chen et al. 2022a, b), (ii) stress and damage (David et al. 1994; Kwon et al. 2001; Lyakhovsky and Hamiel 2007), (iii) equivalent channel concept (Paterson 1983), and (iv) network models (Jivkov et al. 2013; Raoof et al. 2013). Among these four categories, the previous two are the most extensively studied.

Researchers have tried to construct a relationship between hydraulic conductivity and porosity in describing the mechanical compaction. There are two main methods in the existing hydraulic conductivity–porosity relationship: the exponential function model and the power function model. The Kozeny–Carman (KC) model (Carman 1937) is one of the most widely accepted and simple models to describe the variation of hydraulic conductivity K in response to changes in porosity n, which can be expressed as:

$$K = C\left(\frac{g}{\mu_w \rho_w}\right)\frac{n^3}{(1-n)^2 S^2 G_s{}^2} \tag{1.12}$$

where C is an empirical parameter that is influenced by the tortuosity and shape of the flow channels, g is the gravitational constant, μ_w is the dynamic viscosity of water, ρ_w is the mass density of water, and nS and G_s denote the mass-specific surface area and the specific weight of the solid material, respectively.

Several semi-empirical equations (Bernabé et al. 2003; Zhu et al. 2008) and modified KC models have subsequently been proposed to improve the estimation of rock permeability subjected to various loading conditions (Holtz and Kovacs 1981; Costa 2006; Ghabezloo et al. 2009). The KC model assumes that there are no electrochemical reactions between the soil particles and the water, hinting at its weak applicability in clays, although it will work for nonplastic silts (Carrier III 2003). This challenge inspired some researchers, like Li (2020), Chen et al. (2022a, b), etc., to improve the KC model for describing the porosity relationship for clays. Li (2020) presented a modified Kozeny–Carman formula by introducing the pervious void ratio, which accounts for the effective voids that a fluid flows through.

$$k = \frac{C_{KC}}{\alpha\left[\alpha + (\alpha - 1)e\right]^2} \frac{e^3}{(1+e)} \tag{1.13}$$

$$C_{KC} = C_F\left(1/S_0^2\right)(\gamma/\mu) \tag{1.14}$$

where k is permeability, e is void ratio, α is the ratio of the volume of voids to the pervious volume of the soils, γ is unit weight of permeant, μ is viscosity of permeant, C_F is a dimensionless shape constant, with a value about 0.2 (Taylor 1948), and S_0 is specific surface area per unit volume of particles (cm^{-1}).

Based on the effective void ratio e_{eff} and effective specific surface area S_{eff}, Chen et al. (2022a, b) proposed a permeability prediction model modified on Kozeny–Carman as follows:

$$k = C_s \frac{\gamma_p}{\mu\rho_s^2 S_{eff}^2} \frac{e_{eff}^3}{\left(1+e_{eff}\right)} \tag{1.15}$$

where C_s is a coefficient describing the tortuosity of pores, γ_p is the specific gravity, and ρ_s is the density of soil particles.

Another form of the hydraulic conductivity–porosity relationship is expressed by exponential functions. Yang and Aplin (2010) proposed a linear relationship between the log of hydraulic conductivity and the porosity for mudstone over a restricted regime:

$$\ln K = a_K + b_K e + c_K e^{0.5} \tag{1.16}$$

where e is the void ratio, and a_K, $b_{K,}$ and c_K are the coefficients (m^2) that are functions of the clay content. Similar approach referring to the relationship between the hydraulic conductivity and porosity has also been found in other studies (Nagaraj et al. 1994; Dewhurst et al. 1999).

Previous studies have shown that temperature has a significant effect on the hydraulic conductivity of geomaterials (see Sect. 1.2). Hence, the thermal effect on the hydraulic properties should be considered when analyzing the fluid transport laws and assessing the stability of the underground engineering involving high temperatures.

Similar to the Kozeny–Carman equation, Towhata et al. (1993a) suggested that the hydraulic conductivity varies with temperature as:

$$\frac{K^T}{K^{T_0}}=\frac{f(T)}{f(T_0)}\frac{g'(T)}{g'(T_0)}\frac{h'(T)}{h'(T_0)} \tag{1.17}$$

where T_0 denotes the reference temperature, f denotes the effects of pore size and shape, g' denotes the effect of void ratio, and h' denotes the property of pore water at the corresponding temperature. However, Towhata et al. (1993a) assumed the ratio of "f" and "g'" in Eq. (1.2) is equal to unity. That means the change of hydraulic conductivity can be predicted only considering the variation of water properties (viscosity and density) with temperature.

Later, a hydraulic conductivity formula is proposed by Ye et al. (2013) as follows:

$$K(T)=k_{in}\times\rho_w\times g/\mu(T) \tag{1.18}$$

where $K(T)$ is hydraulic conductivity, cm/s; k_{in} is intrinsic permeability, cm^2; ρ_w is water density, g/cm^3; g is gravitational acceleration, m/s^2; and $\mu(T)$ is water viscosity, mPa·s.

Suggesting that the thermal effect on the intrinsic permeability should be considered, the relationship between hydraulic conductivity and temperature was reexpressed by Wang et al. (2017) as follows:

$$K(T)=k_{in}(T)\times\rho_w\times g/\mu(T) \tag{1.19}$$

where $k_{in}(T)$ is the intrinsic permeability, which is dependent on temperatures. According to Wang et al. (2017), $k_{in}(T) = \exp(A + B \times T)$, A and B are constants.

Excavation of underground engineering can lead to a redistribution of internal stress in the host rock, especially in rock with reduced radial stress, where high deviatoric stresses are concentrated. Fractures initiated and developed inside the host rock in the near field, forming a damaged zone and causing changes in its hydraulic properties. It has been observed that stress and damage may influence rock hydraulic properties, especially in these geomaterials with low hydraulic conductivity. Some empirical relationships have been developed to describe the hydraulic conductivity changes based on stress and damage.

David et al. (1994) proposed an exponential permeability–stress relation in the study of the generation and maintenance of pore pressure excess in the crust. The exponential law takes the following form:

$$k=k_0\exp\left[-\xi(\sigma-\sigma_0)\right] \tag{1.20}$$

where k is the permeability at the effective stress σ, k_0 is the permeability at ambient stress σ_0, and ξ is the stress sensitivity coefficient.

Kwon et al. (2001) proposed a power law to describe the permeability of illite-rich shale under different effective stresses:

$$k = k_0 \exp\left[1-\left(\sigma / P_1\right)^m\right]^3 \tag{1.21}$$

where k_0 is the reference permeability at zero effective pressure, σ is the effective stress, P_1 is the effective modulus of the equivalent asperities, and m is a constant between 0 and 1.

Lyakhovsky and Hamiel (2007) proposed a power law that accounts for the variable damage to describe the hydraulic conductivity–stress relationship:

$$K = K_0 \exp\left(\frac{n}{n_0}\right)^z \left(\frac{D}{D_0}\right)^{Z'} \tag{1.22}$$

where D and D_0 are the current damage and the reference damage value, respectively; and Z' is the exponent parameter for damage.

Tang et al. (2002) proposed a coupled flow-stress damage model for rocks by extending Biot's theory to include the effects of stress and damage on permeability:

$$K = \begin{cases} K_0 \exp\left(-\gamma\left(p-\chi p_0\right)\right) & (D=0) \\ \omega K_0 \exp\left(-\gamma\left(p-\chi p_0\right)\right) & (0 < D \leq 1) \end{cases} \tag{1.23}$$

where p is the normal stress; p_0 is the reference normal stress; ω is the damage factor of permeability, which is greater than 1, and it is defined as the increase in permeability caused by damage; and χ and γ are the coefficients defined in Biot's seepage equation and permeability–stress relation, respectively.

Li et al. (2024) developed a novel equation to establish a relationship between effective damage and permeability, accounting for the characteristics of low-permeability non-Darcy flow in rocks.

$$k^* = D^* \frac{\mu \varsigma \Phi}{\lambda\left(1-\Phi\right)\tau} d^2 \tag{1.24}$$

where k^* is the actual permeability, D^* is the effective damage, Φ is the volumetric porosity of rock damage fractures, ς is the drag coefficient particle sphere, λ is the Stocks coefficient, τ is the influence coefficient of adjacent particles, and d is the diameter of resistance spheres.

1.5 Structure and Methodology of the Book

This book contributes to an experimental program focused on both the short- and long-term THM behavior of a clay rock, named Boom Clay, a candidate host rock for nuclear waste disposal in Belgium (Bernier and Neerdael 1996; Li et al. 2010, 2023). On the one hand, based on the short-term experimental results, a transversely isotropic thermo-hydro-mechanical-damage coupled model for clay rock has been proposed and secondarily developed in the finite element code ABAQUS (Abaqus 2006). On the other hand, a thermo-elasto-plastic creep damage constitutive model for clay rock has been developed. Based on the theoretical studies, the proposed models were implemented in the finite element code ABAQUS; then, large numerical simulations were carried out: First, a 3D numerical simulation of hydro-mechanical (HM) perturbation during the connecting gallery construction in HADES was carried out to validate the proposed model. Second, a 3D numerical simulation of PRACLAY in situ heating test in HADES was carried out to assess the THM-coupled response of clay rock during the excavation phase, heating phase, and cooling phase. In detail, this book is organized as follows:

Chapter 1 gives an extensive review of the literature on the topics of THM-coupled behavior of clays.

Chapter 2 presents the experimental program focused on investigating temperature effects on the hydro-mechanical properties of Boom Clay. The program included THM-coupled permeability tests, THM triaxial compression tests, and investigation of microstructural changes using the NMR method. The test apparatus and experimental procedures are presented in detail. The temperature effects on the strength, stiffness, hydraulic conductivity, and microstructure of Boom Clay are analyzed based on the experimental results.

Chapter 3 establishes a transversely isotropic thermo-hydro-mechanical-damage coupled model, which combines the transversely isotropic elastic model with the Drucker–Prager Cap model. The elastic damage evolution, the thermal damage evolution, the plastic damage evolution, the plastic hardening law, and the hydraulic conductivity evolution model are introduced based on the results of the THM-coupled triaxial tests and THM-coupled permeability tests of Boom Clay. The proposed model was developed and implemented in the ABAQUS finite element code through the subroutine USDFLD. Parameter calibration is conducted using the test data obtained in Chap. 2, and a comparison is made between the predictions of the developed model and the experimental data.

Chapter 4 illustrates the experimental activities carried out to investigate the temperature effect on the creep behavior of Boom Clay. Drained triaxial creep tests were performed under well-defined THM loading paths. During the creep stage, the creep load was kept constant while a cyclic temperature process ($40 \rightarrow 60 \rightarrow 80 \rightarrow 60 \rightarrow 40$ °C) was regulated to the temperature variation expected in the real repository. Several aspects were analyzed in detail, including the effect of temperature and time on the creep rate, the irreversibility of the creep behavior after cooling phases, and the mechanism of the THM-coupled creep behavior of Boom Clay.

Chapter 5 first proposes an elasto-plastic creep damage constitutive model based on the overstress theory of Perzyna. Considering the thermal effect, a thermo-elasto-plastic creep damage model is then developed based on the proposed elasto-plastic creep damage model. The temperature effect on the creep hardening behavior was considered. Meanwhile, the thermal damage model is introduced as a further weakening effect on the mechanical behavior due to temperature. The model is implemented in ABAQUS by the CREEP subroutine and USDFLD subroutine.

Chapter 6 shows the results of the 3D numerical simulation of CLIPEX in situ experiments to validate the transversely isotropic thermo-hydro-mechanical-damage coupled model proposed in Chap. 3. The results of the numerical simulation are compared with in situ measurements. The HM coupled transversely isotropic characteristics of Boom Clay are analyzed.

Chapter 7 shows the preliminary results of the large 3D numerical simulation of the PRACLAY in situ heating test, which was carried out to assess the long-term THM-coupled response of Boom Clay. The simulation was carried out based on the damage constitutive model. The excavation disturbance, self-sealing, and temperature effects on the hydraulic conductivity of Boom Clay are considered.

Chapter 2
Experimental Study on THM Behavior of Boom Clay

2.1 Introduction

The temperature of the clay rock barrier is expected to increase after the installation of waste canisters due to the heat emitted by HLW. As aforementioned, the hydro-mechanical behavior of clay rock at high temperatures is quite different from that at ambient temperature (Chen et al. 2024). Thus, the first step for the stability analysis of the radioactive waste repository should consider the temperature effects on the hydro-mechanical properties of clay rock. To understand the thermo-hydro-mechanical properties of the Boom Clay, a series of laboratory tests focused on investigating the short-term influence of increasing temperature on the mechanical and hydraulic properties of Boom Clay were carried out. The testing program included consolidated undrained triaxial compression tests and permeability tests, which were conducted under various stress and thermal conditions. Nuclear magnetic resonance (NMR) tests were also performed at the same time to help understand the mechanism of temperature effects on the hydro-mechanical properties.

2.2 THM Undrained Triaxial Compression Tests

2.2.1 Experimental Preparation

2.2.1.1 Samples Preparation

Boom Clay samples were drilled in cores of natural Boom Clay, which is stored in a humid chamber. These cores were under vacuum in plastic bags initially, but the condition of vacuum could be changed over time. All tests were conducted on samples with "natural" water content (the water content of the blocks) so as not to disturb the samples. To obtain samples with adequate dimensions, cored specimens

W. Chen et al., *Thermo-Hydro-Mechanical Coupling in Clay Rock*,
https://doi.org/10.1007/978-981-95-2562-1_2

were then made by hand trimming (with a knife, see Fig. 2.1). The procedures of hand trimming are difficult because of the macrofissures are clearly induced by drilling process the test samples have to be made in the cores between the macrofissures (Fig. 2.2), mineral fillings, and especially the stiff mineral fillings found in the cored specimens. Macrofissures can be found easily, but as microfissures are concealed, they may be found when the sample has nearly completed. During the hand trimming process, to avoid water loss in the sample surface, the environmental temperature and humidity of the laboratory were strictly controlled. At the same time, samples were made as soon as possible to minimize the sample preparation time to minimize evaporation. The samples can be prepared in less than half an hour.

Eighteen tests were performed on specimens of Boom Clay trimmed from the cores. The axis of the specimens is perpendicular to the bedding. Only 15 representative tests are discussed herein. The initial physical characteristics of the samples were measured during the sample-making process. Water content w was determined on the residues by weighing them on a balance with a precision of 0.01 g. Table 2.1 provides the main characteristics of all the samples tested. It can be observed that the water content is in the known range of $23 < w < 26\%$, and the saturation is in the known range of $91 < S_r < 100\%$. A possible volume measurement of the samples was done by using a precision caliper.

Fig. 2.1 Procedures for making Boom Clay samples for triaxial test

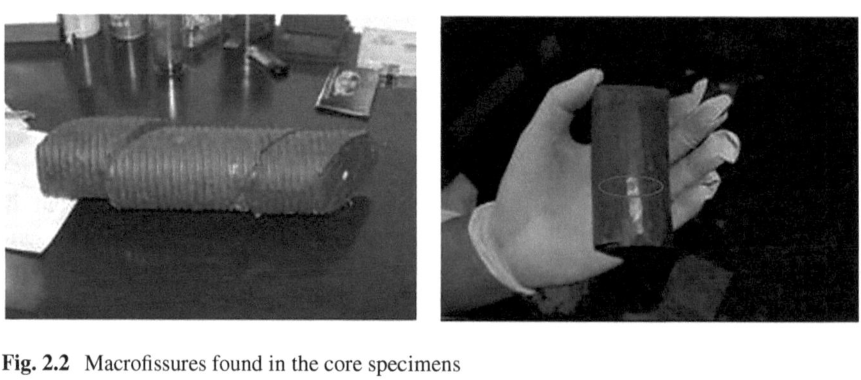

Fig. 2.2 Macrofissures found in the core specimens

2.2.1.2 SBCW Preparation

Synthetic Boom Clay water (SBCW) was used for the THM triaxial compression tests and permeability tests. The compositions of SBCW are essentially salt and $CaCO_3$, and the amounts of salt powder for 1 L of distilled water are provided by EURIDICE (De Craen et al. 2004). After putting the salt into the distilled water, 5 g of $CaCO_3$ was added to the solution. The solution was stirred for 72 h (Fig. 2.3).

Table 2.1 Summary of the THM tests

Test	ρ (g/cm^3)	w (%)	e_0	σ_c (MPa)	p_w (MPa)	T (°C)
HM_{47222}	2.01	24.0	0.697	4.7	1.2	22
THM $_{47401}$	2.01	23.5	0.683			40
THM $_{47402}$	2.01	23.5	0.683			40
THM $_{47602}$	2.04	24.3	0.673			60
THM $_{47801}$	2.01	23.5	0.683			80
HM $_{37221}$	2.03	23.8	0.677	3.7	1.2	22
THM $_{37401}$	2.04	24.2	0.671			40
THM $_{37402}$	2.00	23.5	0.693			40
THM $_{37601}$	2.03	24.0	0.675			60
THM $_{37801}$	2.03	23.8	0.668			80
THM $_{37802}$	2.04	23.5	0.665			80
HM $_{25221}$	2.04	24.7	0.677	2.5	1.0	22
HM $_{25222}$	2.04	24.7	0.677			22
THM $_{27601}$	2.01	23.5	0.683	2.7	1.2	60
THM $_{27801}$	2.01	23.5	0.683			80

Note: ρ is the density, w is the water content, e_0 is the void ratio, σ_c is the isotropic stress, p_w is the back pressure, T is the temperature at which the sample is sheared. Synthetic Boom Clay water was used during the sample saturation

(a) Salt powder provided by EURIDICE (b) Stirred solution

Fig. 2.3 SBCW preparation. (**a**) Salt powder provided by EURIDICE. (**b**) Stirred solution

2.2.1.3 Experimental Device

A double-linkage triaxial testing machine, which is particularly applicable to study the rheological characteristics of soft rock under various THM-coupled conditions, has been designed in the beginning of 2006 and updated 10 years later (Chen et al. 2019) (see Fig. 2.4) (Table 2.2). During the update, the format of the machine has been changed from series to parallel connection to eliminate the influence of the

(a) Initially designed in 2006 (b) Updated in 2016

Fig. 2.4 Double-linkage triaxial testing machine (Chen et al. 2019). (**a**) Initially designed in 2006. (**b**) Updated in 2016

Table 2.2 Main technical index and parameters

Classification	Technical index	Technical parameters (initially designed)	(updated)	Precision (full-scale percentage)
Axial loading system	Maximum force	20 kN	50 kN	≤ ±0.1%
	Displacement	0–500 mm	0–500 mm	≤ ±0.5%
	Axial deformation	0–12(25) mm	0–12(25) mm	± 0.001 mm
	Radial deformation measuring range	0–4 mm	0–4 mm	≤ ±1%
Confining pressure loading system	Maximum confining pressure	5 MPa	10 MPa	≤ ±0.02%
Water pressure loading system	Maximum water pressure	5 MPa	5 MPa	≤ ±0.02%
Temperature controlling system	Maximum temperature	100 °C	100 °C	≤ ±0.5 °C
Others	Sample dimensions	Φ50 × 100 mm and Φ38 × 76 mm choose arbitrary		
	Power	AC 380 V 4 kW		
	Noise	≤ 45 dB		
	Working time	> 2000 h		

weight of the upper chamber for the series connection. Several technical parameters were also updated. Put aside the connection format, this machine can automatically stabilize voltage and record stress–strain variation curves and temperature duration curves using the technology of servocontrol, ball screw, confining pressure, water pressure, thermal loading, and LVDT deformation sensors. All the load cells, hydraulic pressure transducers, and LVDT deformation sensors can work stably and accurately at higher temperatures. In addition to the function of an ordinary triaxial device, because of the unique design form of this machine, it can also simultaneously test two samples with the same confining pressures and water pressure, but different axial pressures and temperatures. The service conditions show that it is a stable machine with complete functions, high precision, and, most importantly, high efficiency.

The Components of the Instrument

The schematic diagram of the double-linkage triaxial testing machine is shown in Fig. 2.5. The device consists of two triaxial cells, two LVDT deformation measurement systems, two axial loading systems, two temperature controlling systems, one confining pressure loading system, one water pressure loading system, and one data logger system. This allows the machine to simultaneously test two samples.

The axial load is servo-controlled by the axial loading frame through the ball screw and measured through a load cell (HBM U10M, see Fig. 2.6). The measuring range of the load cell is from 0 to 50 kN. The measurement precision of the sensor is better than 0.04%, with a range from −30 to 85 °C of temperature compensation.

The confining pressure and the water pressure are controlled by two pressure generators and measured through hydraulic pressure transducers. Hydraulic oil is used as the cell fluid, and the confining pressure is measured by two hydraulic pressure transducers (GE DRUCK UNIK 5010, see Fig. 2.7). SBCW is used for water pressure loading. Two stainless steel ducts are connected to the top cap and the base of the sample to allow water pressure loading to the sample by flushing. The water pressure of the pressure generators, top and base of the sample, is measured simultaneously by three hydraulic pressure transducers (also GE DRUCK UNIK 5010). The measuring range of the hydraulic pressure transducer is from 0 to 5 MPa. The measurement precision of the sensor is better than 0.1%, with a range from −40 to 125 °C of temperature compensation.

The temperature controlling system consists of the heating coil, thermocouple, and temperature controller. Temperature is measured simultaneously by a thermocouple submerged in the cell fluid and a thermocouple placed in the heating coil. The power supplied to the coil can be automatically adjusted by the temperature controller. The heating rate can be linearly controlled by regulating the current. The temperature controller and a heating–cooling cycle example (with the rate of 0.3 °C/h) controlled by it are shown in Figs. 2.8 and 2.9, respectively.

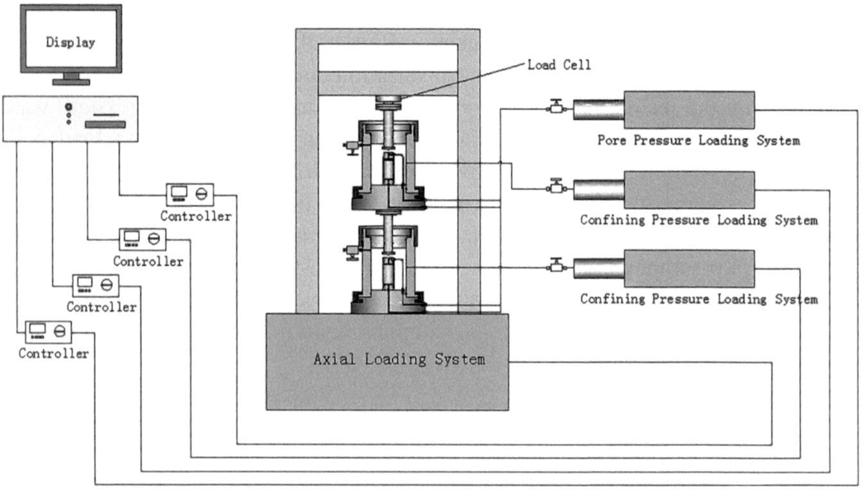

(a) Initially designed in 2006

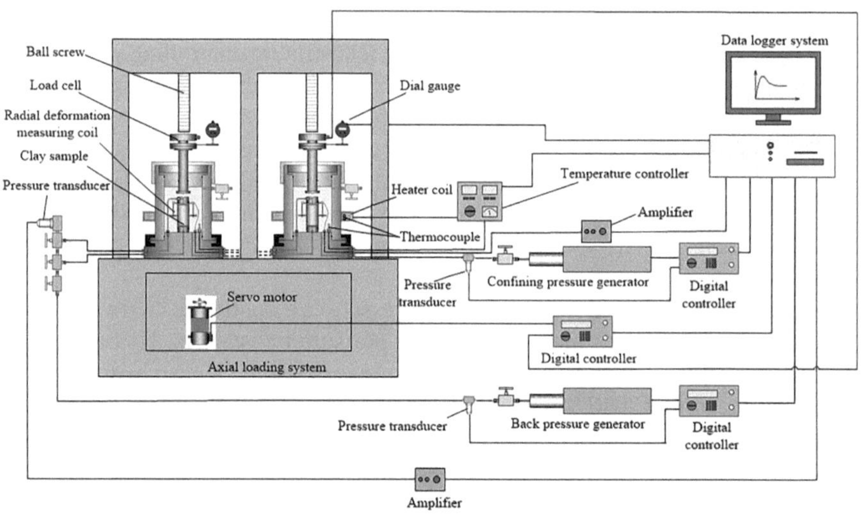

(b) Updated in 2016

Fig. 2.5 Schematic diagram of double-linkage triaxial testing machine. (**a**) Initially designed in 2006. (**b**) Updated in 2016

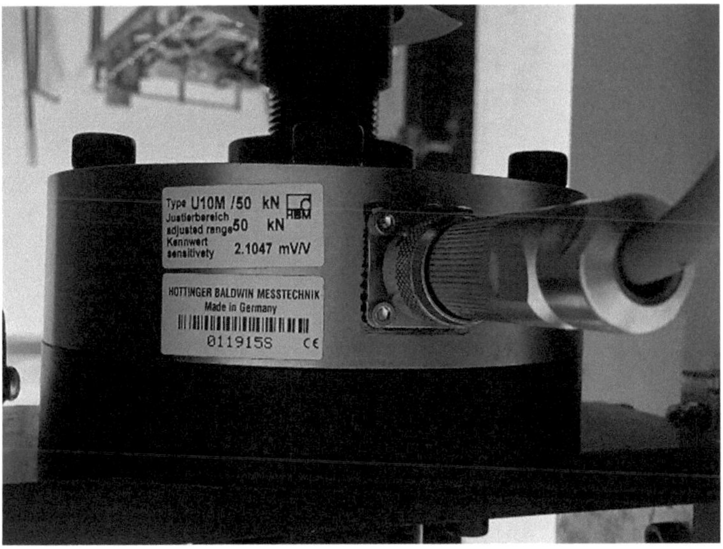

Fig. 2.6 HBM U10M/50 KN load cell

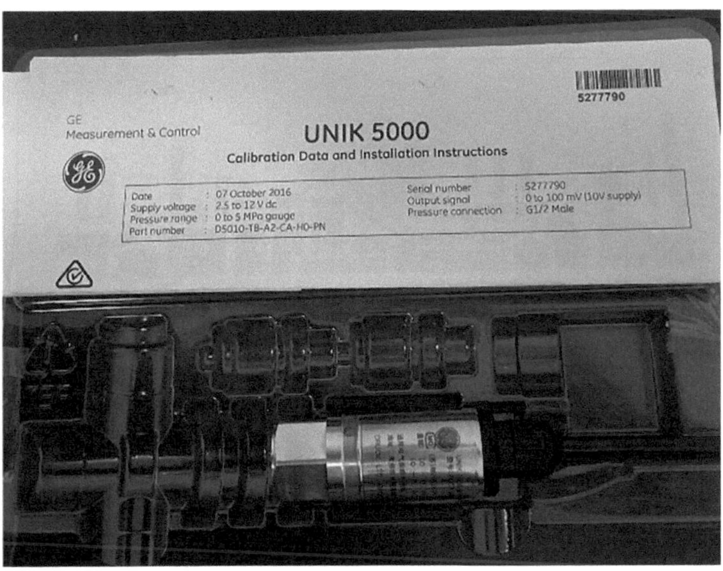

Fig. 2.7 GE DRUCK UNIK 5010 hydraulic pressure transducer

Fig. 2.8 Temperature controller

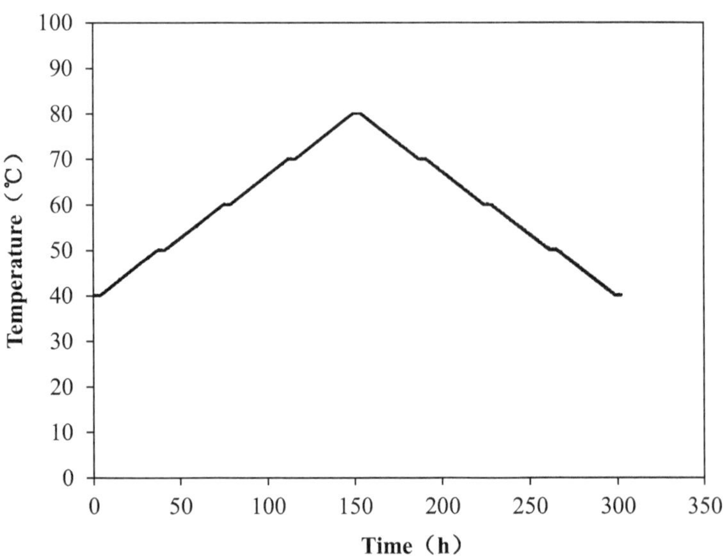

Fig. 2.9 A heating–cooling cycle controlled by a temperature controlling system with a rate 0.3 °C/h.

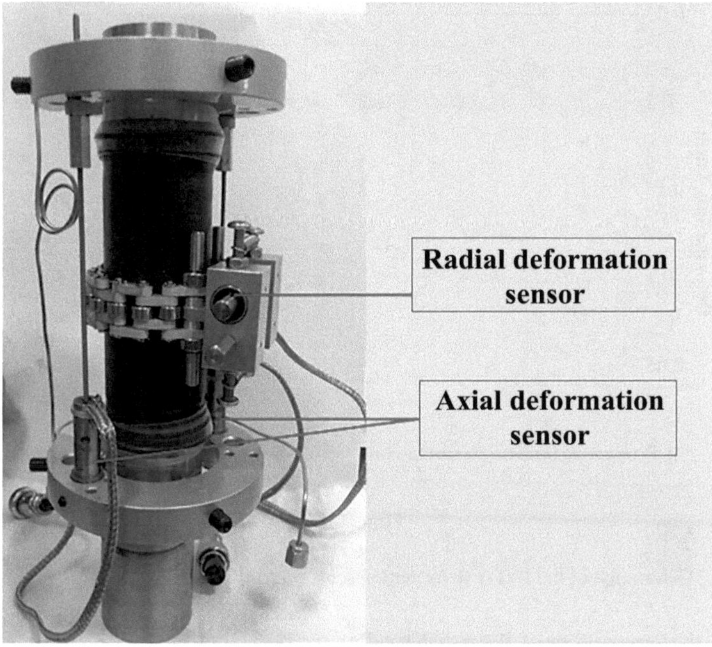

Fig. 2.10 Axial and radial LVDT deformation sensors

The axial and radial deformations are monitored by the LVDT deformation measurement system. The measuring range of the axial and radial LVDT deformation measurement system is from 0 to 10 mm, and the measurement precision of the sensors (Fig. 2.10) is better than 0.1%. The temperature effect on the axial and radial LVDT deformation measurement system can be calibrated accurately. Careful calibration of the LVDT deformation sensors during a heating–cooling cycle was made on a steel sample (Fig. 2.11).

The Advantages of the Instrument

The advantages of the double-linkage triaxial testing machine are as follows:

(1) The high efficiency machine can simultaneously test two samples with the same confining pressures and water pressure, but different axial pressures and temperatures.
(2) The machine enables sample re-saturation under certain effective stress, THM-coupled permeability test (both transient and constant-head method), short-term THM-coupled test (such as drained and undrained triaxial compression tests), and long-term THM-coupled test (such as drained and undrained creep tests)

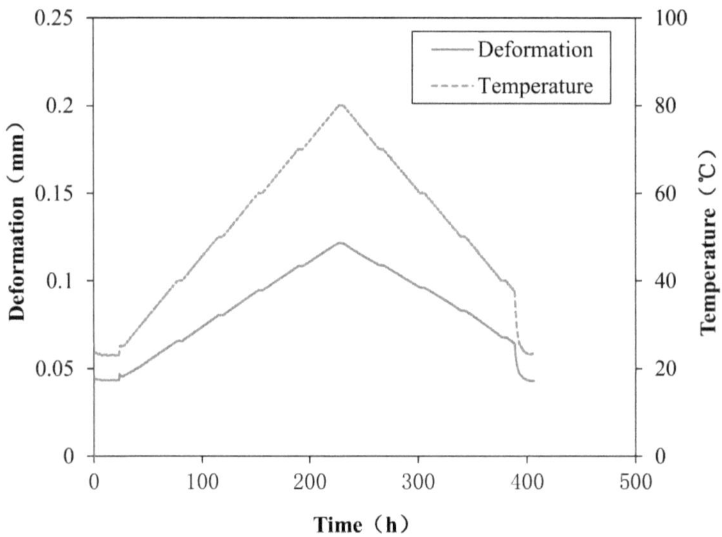

Fig. 2.11 Calibration of the LVDT deformation sensor during a heating–cooling cycle

(3) All the controllers of the axial loading system, confining pressure, and water pressure loading system are DOLI EDC 220, which results in a desirable characteristic of less load fluctuation.
(4) The machine is equipped with a UPS uninterrupted power supply system, which can guarantee the machine works uninterruptedly under outage conditions.

2.2.1.4 Test Procedures

The low permeability makes the mechanical behavior of Boom Clay a complex one. The tests are time-consuming as the test procedures must carefully take account of the specimen saturation, the temperature increasing rate, good control of the axial stress, confining pressure, and pore water pressure (Fig. 2.12). To restore the in situ states of Boom Clay as much as possible, the isotropic stress of each sample was first increased to in situ mean effective stress 2.5 MPa before any contact with water. This procedure reduced the swelling and thus the related microstructure changes (Delage et al. 2007; Coll 2005; Cui et al. 2009). Specimen saturation was achieved by increasing pore water pressure at the two ends (top and bottom) of the specimen. In detail, the soil sample was saturated with synthetic water of ionic composition similar to the in situ pore water under the in situ mean effective stress of 2.5 MPa. During the saturation stage, the isotropic stress and the back pressure were increased step-by-step by keeping the constant mean effective stress of 2.5 MPa, until a back pressure of 1.2 MPa. The loading rates for both isotropic stress and back pressure (as indicated in Fig. 2.12) were considered to be slow enough to ensure complete drainage during loading. Then, the specimens were consolidated to a target mean

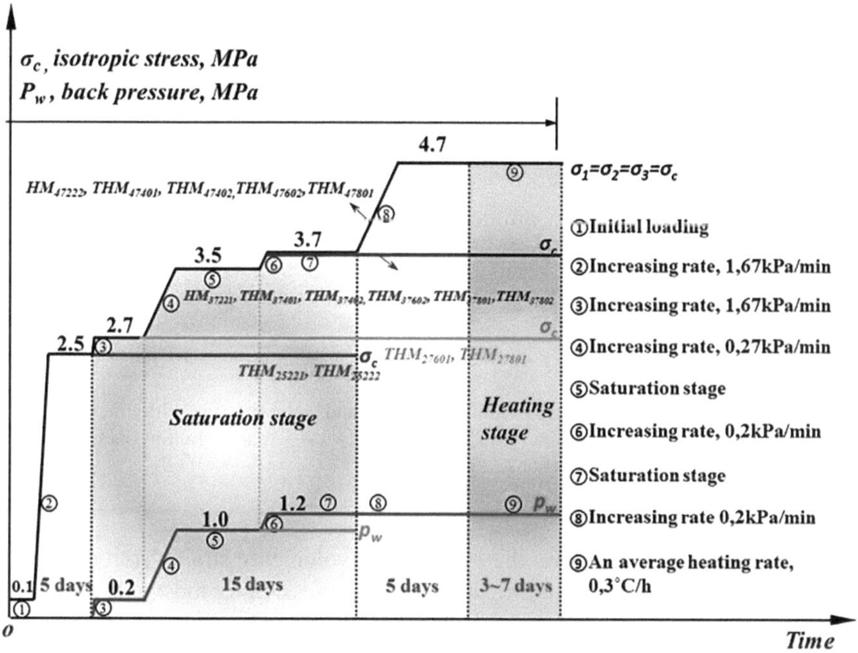

Fig. 2.12 Load procedures for saturation and temperature elevation

effective stress under the same back pressure of 1.2 MPa after ensuring saturation (by checking the Skempton coefficient, the value of which was observed to be larger than 0.9 for most of the samples). Afterward, the specimens were heated with an average heating rate of 0.3 °C/h. This rate was confirmed to be slow enough to ensure a constant mean effective stress inside the specimens by Sultan et al. (2002) and Delage et al. (2000). Finally, after stabilization of temperature in the cell, the specimens were sheared with the compression rate of 20 μm/min under constant temperatures. For most of the tests, the deviatoric phase lasted for about 12 h. It has to be mentioned that several samples, including THM_{25221}, THM_{25222}, THM_{27601}, and THM_{27801}, were saturated under the mean effective stress of 1.5 MPa because of the low confined pressure.

2.2.2 *Test Results*

2.2.2.1 Volume Change Behavior During Isotropic Compression

All the triaxial tests were first submitted to isotropic compression. For the triaxial tests under high temperatures, one more step of temperature elevation was performed before the deviatoric phase. Figure 2.12 shows the stress path followed before the deviatoric phase for each test, including the load increasing rate. As a

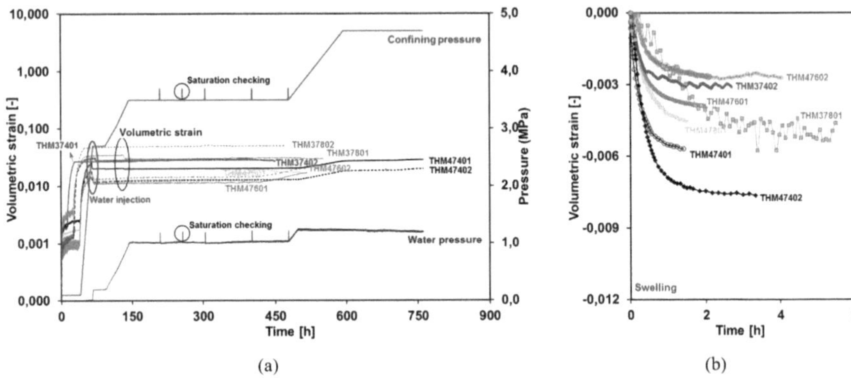

Fig. 2.13 Evolution of the volumetric strain, confining and water pressure with time during isotropic compression. (**a**) Overall view. (**b**) Local enlarged view—water injection (Reset the volumetric strain before the water injection)

representative, Fig. 2.13 shows the evolution of volumetric strain, isotropic stress, and water pressure with time during isotropic compression for most of the samples. Note that volume changes were calculated from the measurements of axial and radial strain. Common results are identified before water injection: contraction occurred during the increase in the mean effective stress. Indeed, the re-saturation procedure reduced the clay swelling. However, as shown in Fig. 2.13b, the swelling caused by water injection is not eliminated. The magnitude of the swelling, as well as the total deformation during saturation, is not the same for all the samples. In 2–4 h after contacting with water, the volumetric strain due to swelling is in the range from 0.3% to 0.8%. Overall, the difference in swelling or total deformation is attributed to the different initial water content. The higher the initial water content, the smaller the swelling during the water injection, and vice versa. This observation was also recognized by Cui et al. (2009) and Delage et al. (2007). According to Cui et al. (2009), the volumetric strain due to swelling decreased from 2.3% to 0.8% (in 100–400 h) when the water content increased from 19.5% to 21.6%. Soil drying, storage, trimming process, the inevitable difference between laboratory conditions and in situ, high content of clay minerals, etc. may be responsible for the swelling observed during the water injection.

2.2.2.2 Thermal Volume Change Behavior During Temperature Elevation

Figure 2.14 shows the typical thermal volume changes versus time during the heating stage. It can be observed that there is a slight expansion before the time around 90 h, when the temperature was increased to about 49 °C, which corresponds to the temperature at which the volumetric expansion/contraction transition occurs for the soil at OCR = 2 (see Fig. 2.15), and larger contraction follows afterward. The detailed discussion will be shown later. A final contraction is observed for most of the tests.

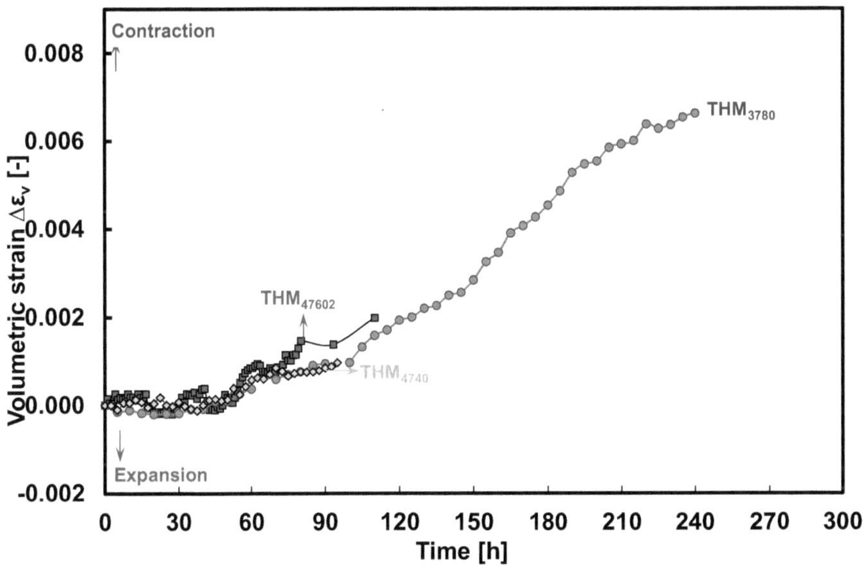

Fig. 2.14 Evolution of the volumetric strain with time during the heating stage

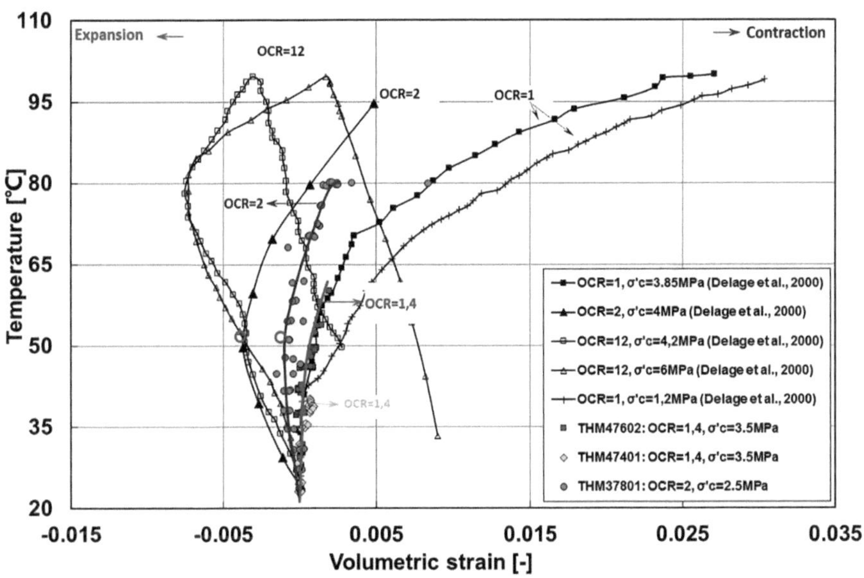

Fig. 2.15 Comparison of the thermal volumetric changes of Boom Clay at different OCRs between Delage et al. (2000) and this study

Previous studies have shown that the overconsolidation ratio (OCR) has a significant effect on the contracting/expanding nature of the volume change of a sample heated under a constant load (Delage et al. 2000, 2004; Sultan et al. 2002; Cui et al. 2009; Plum and Esrig 1969; Baldi et al. 1988; Towhata et al. 1993a; Paaswell 1967). To identify the effect of temperature on the thermal volume changes of Boom Clay, Sultan et al. (2002) and Delage et al. (2000, 2004) performed a series of thermal consolidation tests at various OCRs (OCR = 1, 2, and 12) and temperatures (22–100 °C) on intact Boom Clay samples. Their results highlight that OCR controls the nature of volume changes of Boom Clay. At low OCRs close to one, only plastic contraction is observed, whereas elastic expansion occurs before plastic contraction at higher OCRs. Thermal contraction decreases with the increasing OCR. This observation was further verified by Cui et al. (2009) later.

Figure 2.15 presents the typical thermal volumetric changes of Boom Clay at different OCRs, together with the study of Delage et al. (2000). Note that, in our study, the clay OCR is defined in terms of isotropic stresses instead of the commonly used vertical stresses in soil mechanics here.

$$\mathrm{OCR} = p_c' / p' \tag{2.1}$$

where p'_c is the preconsolidation pressure. According to Bésuelle et al. (2014), Yu et al. (2012), and Deng et al. (2011), p'_c is taken to be 5 MPa. p' is the mean effective stress, $p' = \sigma'_c$ (σ'_c is effective isotropic stress).

In this study, the samples underwent OCR = 1.4, which showed only plastic contraction during the heating phase. This is in good accordance with the results of Cui et al. (2009), in which they proved that only contraction and thermal consolidation were observed at OCR = 1.4 and 1.25. Moreover, this study further confirms that a larger contraction takes place after an initially smaller thermal expansion at OCR = 2. The temperature at which the thermal volume change behavior passes from expansion to contraction is about 50 °C at OCR = 2. This value is well consistent with the study of Sultan et al. (2002) and Delage et al. (2000, 2004), in which they proved that the temperature at which the expansion/contraction transition increases with the increasing OCR (T = 80 °C at OCR = 12, T = 50 °C at OCR = 2). Similar results were obtained by Baldi et al. (1988) through performing heating tests at constant effective pressures and isotropic loading tests at various constant temperatures on low-porosity clays, which showed that the temperature at which the transition between contraction and expansion occurred was increasing with OCR (T = 58 °C at OCR = 2.5).

For clayey soils, the thermal contraction can be related to physicochemical clay–water interactions based on the change in thickness of the double layer with an increase in temperature by various authors, such as Plum and Esrig (1969), Baldi et al. (1988), Towhata et al. (1993b), etc.

2.2.2.3 Deviatoric Phase

Stress–Strain Behavior

Figure 2.16 illustrates the test results related to the deviatoric phase in terms of variations of deviatoric stress versus axial or radial strain, from which one can recognize that the Boom Clay exhibits the typical elasto-plastic behavior. The stress–strain curves start with an approximate linear stage until yield occurs at stress σ_y (see Fig. 2.16a, take THM_{47801} as an example), where an obvious deviation from this linear part appears. Ideally, the yield stress σ_y is the boundary point between the elastic and plastic regions. Strictly speaking, the elastic region here is a nominal one, as the existing experimental tests (Rousset 1988; Baldi et al. 1991) have shown that the hysteretic unloading–reloading loops occurred at very low strain, which means that certain irreversibility took place, and the pure elastic stage is very limited: here the stress σ_e is the limit of linear proportionality. According to Popp and Salzer (2007) and Zhang (2016), the yield point indicates the onset of microcracking of Opalinus clay, which was detected through the shear wave velocity. Unfortunately, though some shear wave velocity measurements have been done on Boom Clay (Neerdael et al. 1992; Piriyakul and Haegeman 2009) for a different purpose, no evidence was found that the opening of the microfractures occurs beyond the yield point based on these measurements. The yielding of Boom Clay can be linked to the modification of the microstructure or the microfissures occurring. After the yield strength, the Boom Clay enters the plastic deformation region. Further increasing the stress results in a nonlinear plastic hardening process until it reaches the peak stress σ_p, known as peak shear strength. There is a peak in the stress–strain curve for all the samples. From the peak point, the shear resistance reduces with the continuous development of plastic shear strains. The stress–strain curve "turns down." A strain-softening behavior occurs. The same form of softening response has been observed in Boom Clay by many other authors (Coll 2005; De Bruyn and Thimus 1996; Horseman et al. 1987, 1993). Finally, the plastic shear strain continues gradually with a plateau value of shear resistance until it reaches the residual shear strength (σ_r).

Due to the differential thermal expansion coefficients of the pore fluid and the clay particles, the saturated clays with low permeability react to the temperature increase with a rise in pore water pressures. In turn, the increase in pore pressure induces a reduction of the mean effective stress, which may lead to shear failure or hydraulic fracturing under certain conditions. Figure 2.17 presents the excess pore pressure as a function of strain at different initial mean effective stresses p'_0 and temperatures. As can be seen from the figure, heating induces a large pore pressure. Under the p'_0 of 3.5 MPa (Fig. 2.17a), the pore pressure buildup increases from about 0.85 MPa to 1.65 MPa when temperature increases from 22 °C to

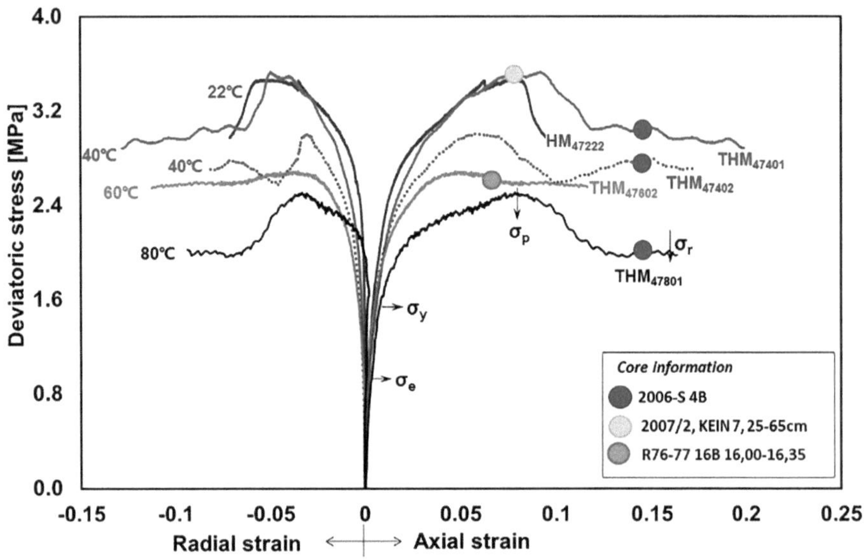

(a) p'_0= 3.5 MPa

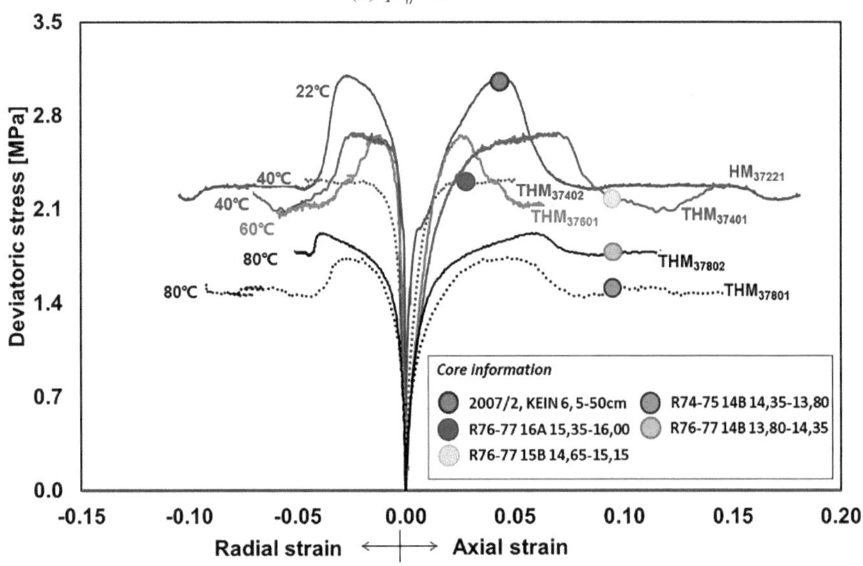

(b) p'_0= 2.5 MPa

Fig. 2.16 Deviatoric stress as a function of strain. (**a**) p'_0 = 3.5 MPa; (**b**) p'_0 = 2.5 MPa; (**c**) p'_0 = 1.5 MPa

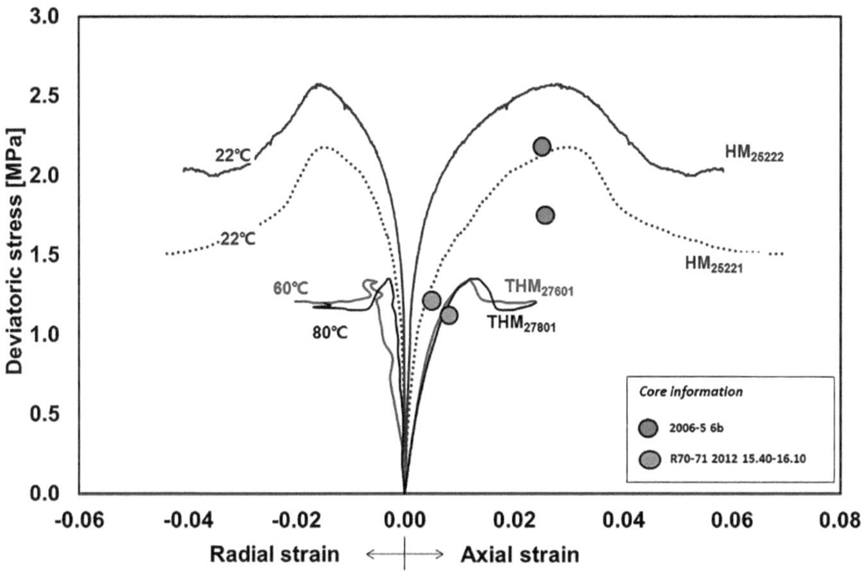

(c) p'_0=1.5 MPa

Fig. 2.16 (continued)

80 °C. Accordingly, the mean effective stress decreases, and thus, the mechanical behavior of Boom Clay changes. Besides, the initial mean effective stress $p'0$ also has an impact on the excess pore pressure. The compression under high p'_0 gives rise to high excess pore pressure under the same thermal condition. Take the excess pore pressure variation at the temperature of 80 °C for instance, the value of the highest excess pore pressure increased from 0.6 MPa to 1.6 MPa when $p'0$ increased from 1.5 MPa to 3.5 MPa.

Stress (σ_p, σ_y, σ_r)—Temperature Relation

Figure 2.18 presents (peak stress σ_p, yield stress σ_y, and residual stress σ_r) as a function of temperature at different initial mean effective stresses p'_0. It shows that the shear behavior of Boom Clay strongly depends on the level of the initial mean effective stress p'_0 (or the level of overconsolidation, OCR = 1.4, 2.0, 3.3) and the thermal condition. Under the same thermal condition, the higher the initial mean effective stress p'_0, the larger the shear strength σ_p. Take the tests at a temperature of 22 °C as an example: σ_p decreases from about 3.4 MPa to 2.3 MPa when p'_0 decreases from 3.5 MPa to 1.5 MPa. Similar results were obtained by Le (2008) and Coll (2005).

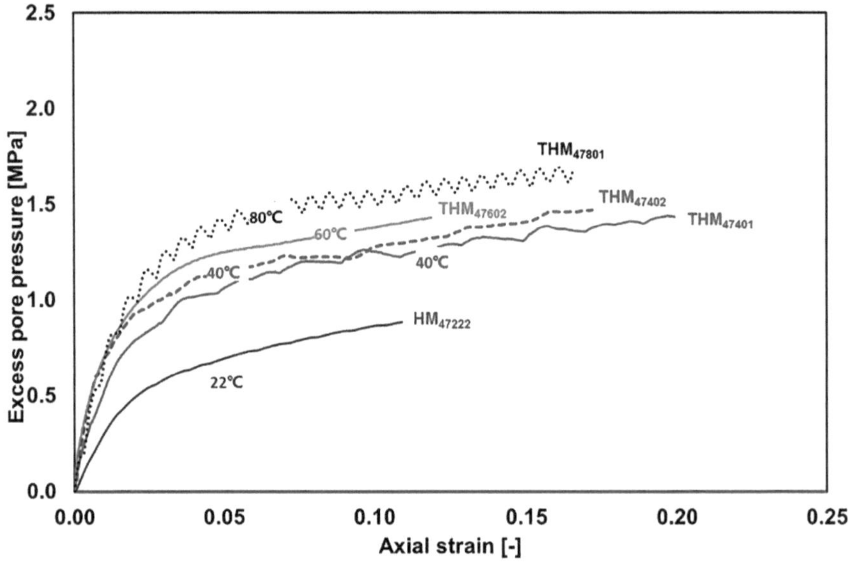

(a) p'_0= 3.5MPa

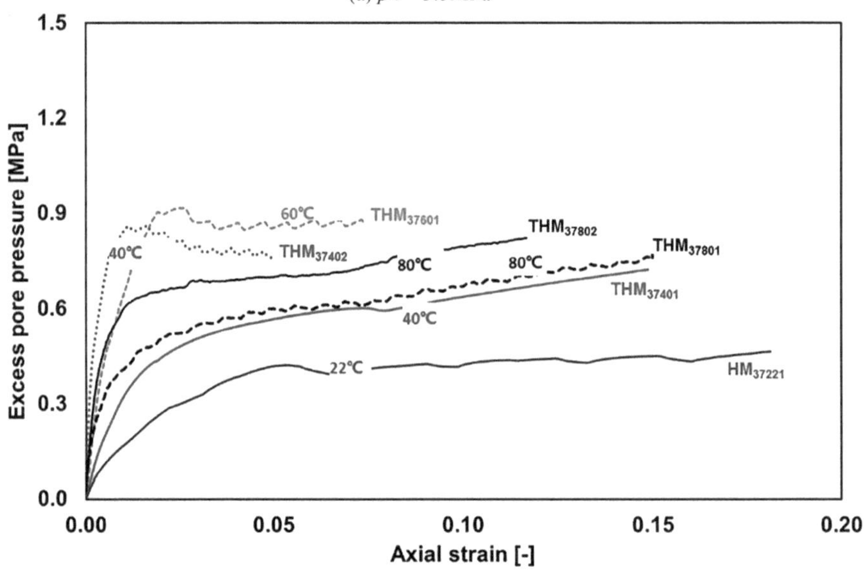

(b) p'_0=2.5 MPa

Fig. 2.17 Excess pore pressure as a function of strain at different initial mean effective stresses p'_0 and temperatures. (**a**) p'_0 = 3.5 MPa; (**b**) p'_0 = 2.5 MPa; (**c**) p'_0 = 1.5 MPa

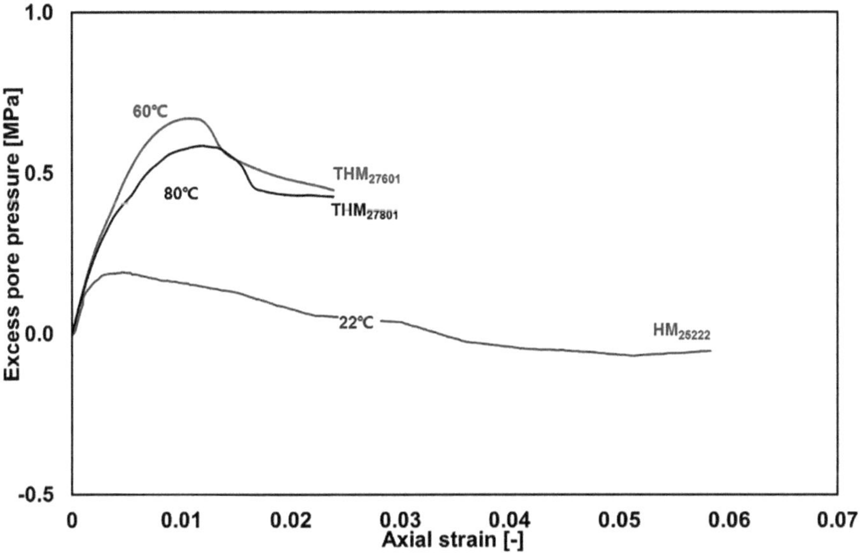

(c) p'_0=1.5 MPa

Fig. 2.17 (continued)

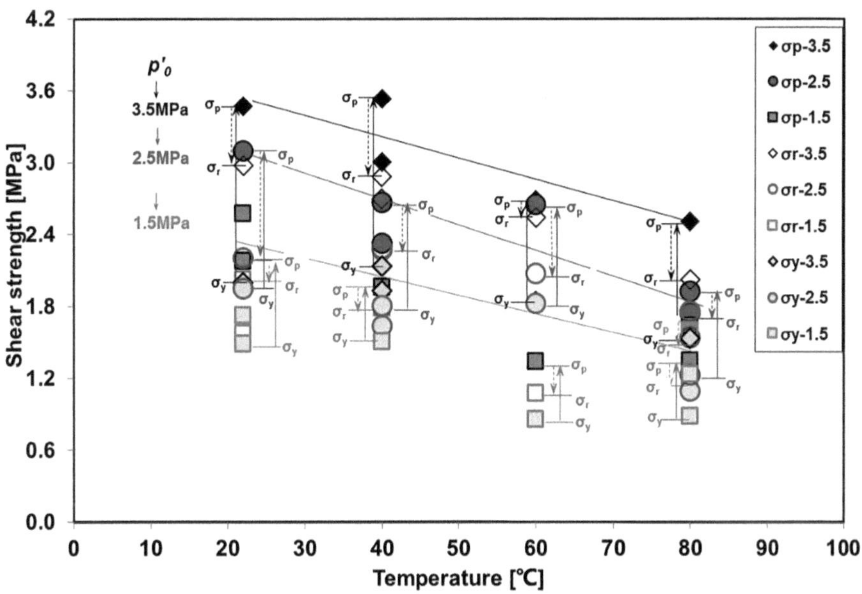

Fig. 2.18 Peak stress σ_p, yield stress σ_y and residual stress σ_r as a function of temperature at different initial mean effective stresses p'_0

Le (2008) performed triaxial drained tests on intact Boom Clay samples under the compression rate of 1 μm/min at the room temperature of 25 °C and found that σ_p decreases from about 3.0 MPa (sample E10) to 2.25 MPa (sample E09) when p'_0 decreases from 3.5 MPa to 1.5 MPa. Conversely, Coll (2005) performed similar drained triaxial tests on intact Boom Clay samples under the compression rate of 0.25 μm/min at the room temperature of 25 °C. They found that σ_p decreases from about 2.4 MPa (sample BC20) to 0.7 MPa (sample BC11) when p'_0 decreases from 2.3 MPa to 0.4 MPa. Meanwhile, the temperature elevation significantly weakened the shear resistance of the sample. Under the initial mean effective stress of 3.5 MPa, σ_p decreases from 3.4 MPa to 2.5 MPa when the temperature elevates from 22 °C to 80 °C, resulting in about a 25% reduction upon an increase of 58 °C in temperature. This result is consistent with those obtained by Le (2008) and De Bruyn and Thimus (1996) showed that σ_p decreased from 3.3 MPa (Sample A-03) to 2.4 MPa (Sample A-10) when the temperature increased from room temperature to 80 °C (with the test conditions: $p'_0 = 3.0$ MPa, compression rate 2.2×10^{-6} s^{-1}, drained condition). The thermally induced strength weakening is also found in other clayey soils, such as Callovo-Oxfordian claystone and Opalinus clay (Zhang et al. 2007; Zhang 2016).

Elastic Modulus–Temperature Relation

The initial secant elastic modulus is computed from the slope of the quasi-linear proportion of the stress–strain curves at different temperatures. Figure 2.19 reports the elastic modulus E_v as a function of temperature at different initial mean effective stress, together with the results of Le (2008) and Coll (2005). It can be seen that,

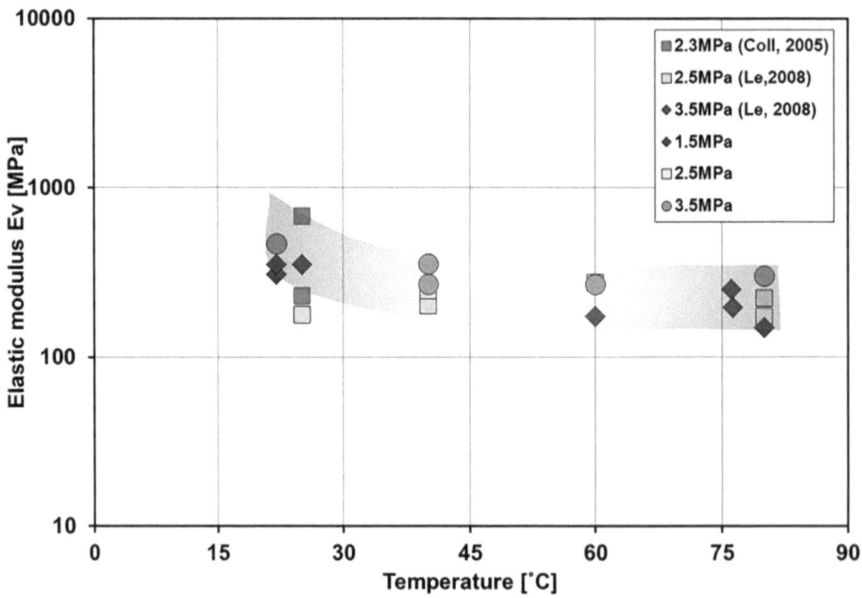

Fig. 2.19 The elastic modulus as a function of temperature from literature and this study

generally speaking, E_v decreases with increasing temperature. The decreasing trend becomes slower and slower as temperature increases. Take the tests under the initial mean effective stress of 3.5 MPa, for instance, E_v decreases from about 400 MPa to 200 MPa when the temperature increases from 22 °C to 80 °C. This indicates that the clayey soil becomes less stiff with the increasing temperature. This may explain the phenomenon of the shear strength declining during temperature elevation. Similar results were obtained by Le (2008) by performing similar tests on intact Boom Clay samples, in which the secant elastic modulus was found to be decreased from about 350 MPa (sample E10) to 197 MPa (sample E15) when temperature increased from 25 °C to 76.1 °C under the initial mean effective stress of 3.5 MPa. It can also be observed that the results of Le (2008) and Coll (2005) (tangent elastic modulus) agree with those obtained by this study. This agreement also indicates that the elastic region of Boom Clay is very limited. Note that values for elastic modulus may be derived from different methodologies (secant versus tangent modulus), accounting for some of the observation divergences.

2.2.3 *Discussions*

2.2.3.1 Failure Envelope in p'-q Plane

An increase in temperature significantly influences the hydraulic and mechanical behavior of clayey soils and some material properties, such as strength, friction angle, and cohesion. But what makes the material properties, such as cohesion and friction angle, of some clayey soils sensitive to temperature increase? No direct experimental evidence is available at present to answer this question. Maybe it is caused by the differential response of water adsorbed on clay to physicochemical changes. An elevated temperature of a clay sample gives rise to the thermal expansion of the pore water and the soil solids, increasing pore water pressure, a decrease in the viscosity of the water, and an increase in the thermal energy, causing self-diffusion of the bound water. In undrained conditions, due to the buildup of thermal pore pressure, bonds carrying stress will tend to relax through diffusion of the bound water, thus throwing some load onto the pore water, resulting in a reduction of the effective stress. A thermally induced failure was evidenced by Menaceur et al. (2016) through performing undrained heating tests on Callovo-Oxfordian claystone.

Regarding the thermal effect on shear strength, contradictory results were found from the literature review. It shows that shear strength can decrease (Le 2008; Hueckel and Baldi 1990; Cekerevac and Laloui 2004; Graham et al. 2001), or stay constant (Burghignoli et al. 2000) with the elevated temperature. While Cui et al. (2000) and Sultan et al. (2002) argue that the effects of temperature on the shear strength of clay are strongly dependent on the volume change induced by heating. It was further confirmed by Hong et al. (2013) recently.

The strength envelope at the critical state of Boom Clay at ambient temperature reported by various authors is shown in Fig. 2.20. As the data is scattered, people tend to have different points of view regarding this issue. If using a straight line to

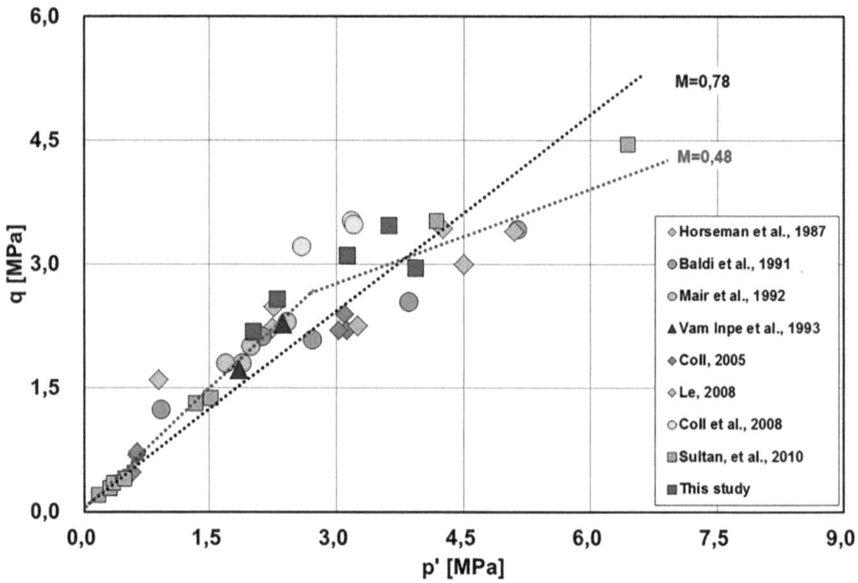

Fig. 2.20 Failure envelopes in p'-q plane for Boom Clay under room temperature

depict the relationship between the mean effective stress and deviatoric stress in the critical state, the slope of the CSL is about 0.78. However, a bilinear relationship can also be adopted according to Deng et al. (2011). The intersection point, i.e., corresponding to approximately the in situ effective stress, separates the envelope into two linear relations: the low stress and the high stress part. The part of high stress is linear with a slope of 0.48. Based on these data, we can obtain a global friction angle and cohesion, about 16° and 0.31 MPa, respectively.

Figure 2.21 reports the failure envelopes in p'-q plane for Boom Clay at different temperatures obtained by this study. The temperature effect on the slope of CSL M is not so evident (it seems to increase slightly from 0.78 to 0.8 when the temperature increases from ambient temperature to 80 °C). There are, however, not enough test data to establish a clear variation law of the slope of CSL with the temperature. Figure 2.22 presents the thermal effect on the slope of CSL, together with that obtained from the literature. Temperature elevation can cause the slope of CSL to increase slightly (Hueckel and Pellegrini 1989), decrease (Robinet et al. 1996), or remain unchanged (Hueckel and Baldi 1990; Graham et al. 2001; Burghignoli et al. 2000; Cekerevac 2003). According to Hueckel and Baldi (1990), the slope of CSL M for Boom Clay in ambient and elevated temperatures is 0.82 and 0.88, respectively. The slope of CSL (M) is increased by about 0.06 as the temperature is elevated to about 68 °C.

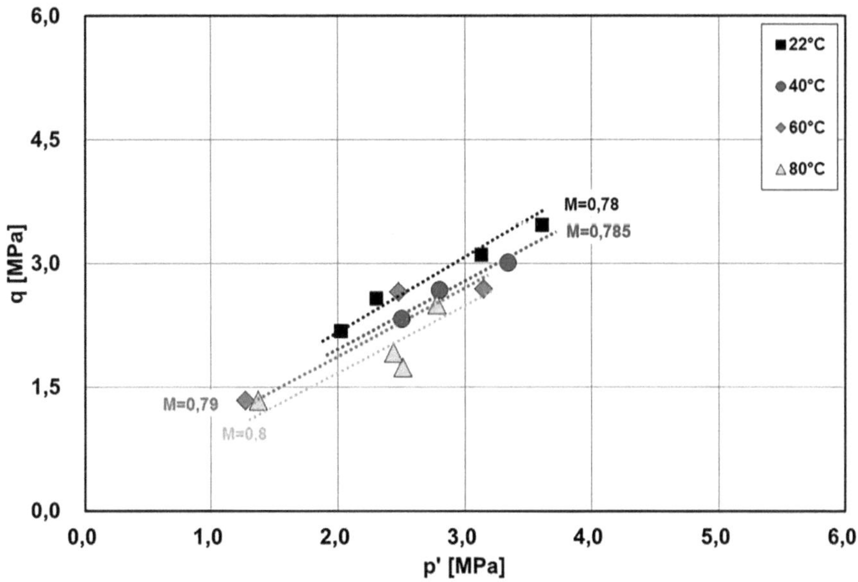

Fig. 2.21 Failure envelopes in p'-q plane for Boom Clay at different temperatures

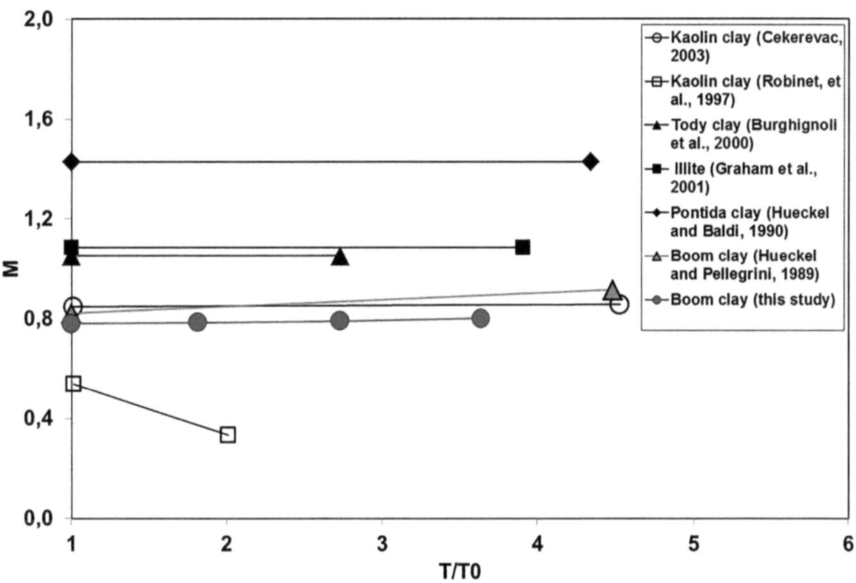

Fig. 2.22 Effect of temperature on the slope of CSL

2.2.3.2 Thermal Effect on Cohesion and Friction Angle

Fitting a Mohr-Coulomb envelope to test data, we can get the cohesion and friction angle of Boom Clay. The cohesion *c*, friction angle φ, or residual cohesion c_r, residual friction angle φ_r were determined by the peak or residual stress values (see Fig. 2.16) and shown in Fig. 2.23. One can observe that temperature elevation has a significant impact on cohesion. The cohesion decreases from about 0.4 MPa to 0.13 MPa when temperature increases from 22 °C to 80 °C. Heating weakens the bonds between the interparticles, and thus, the cohesion decreases. The weakening of cohesion during heating may induce less stiffness and thus lower the shear strength of Boom Clay. However, the trend of the thermal effect on the friction angle is not clear. As can be seen from Fig. 2.23, the friction angle for Boom Clay is around 18°. The values of both the friction angle and cohesion here show good agreement with the literature (Coll 2005; Lima, 2011; Baldi et al. 1991; Horseman et al. 1987), in which the friction angle of Boom Clay is found to be in the range of 13°–25°, and the cohesion in the range of 0.1–0.5 MPa. Comparing the parameters at peak with those residual ones, it can be seen that the friction angle changes a little when it reaches the residual state. But the cohesion decreases at the residual state. At the temperature of 80 °C, the cohesion decreases from about 0.13 MPa at the peak to 0.05 MPa at the residual state. It is easy to understand this weakening of cohesion. Breaking of bonds would occur to permit interparticle reorientation during compression. Therefore, at the residual state, the clay particles have already been restructured, and hence, the interparticle bonds are probably weakened.

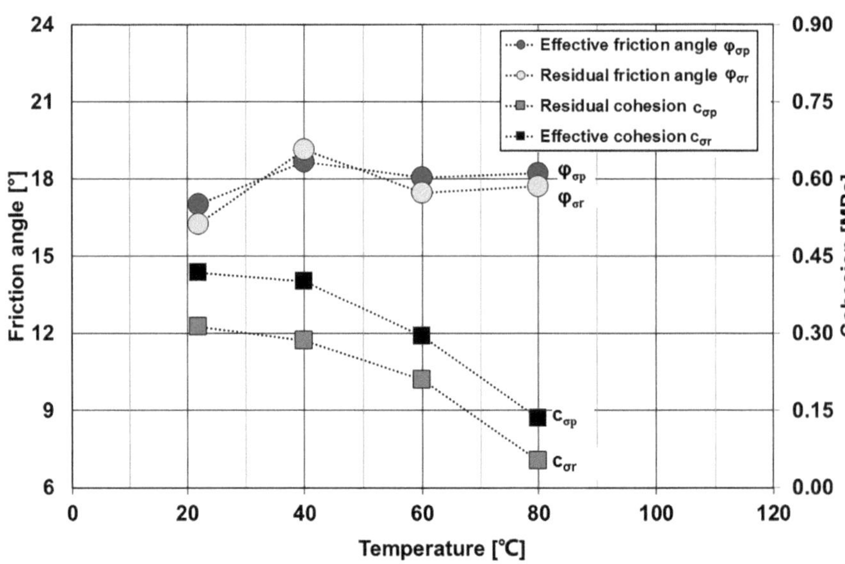

Fig. 2.23 Evolution of the effective cohesion and friction angle with temperature

Fig. 2.24 Typical patterns of shear planes observed on the samples after failure

The results show good repeatability in terms of the failure mode of all the samples. A typical failure sample after the test was shown in Fig. 2.24. It can be seen that the inclination of the plane of rupture θ is about 53.5°. Thus, we can preliminarily deduce that the friction angle of Boom Clay is about 17° from the relation $\theta = 0.5\varphi + 45°$, which is close to the friction angle obtained through calculation shown in Fig. 2.21.

To better analyze the thermal effect on clay friction angle, the results of the present work are plotted together with those obtained from other works in the same figure (respectively, Fig. 2.25). Literature data show that an increase in temperature can cause a higher (Burghignoli et al. 2000; Hueckel and Pellegrini 1989), lower (Robinet et al. 1996; Ghahremannejad 2003), or unchanged (Hueckel and Baldi 1990; Cekerevac and Laloui 2004; Graham et al. 2001; Burghignoli et al. 2000; Yavari et al. 2016) friction angle.

Note that the thermal effect on the cohesion of geomaterials in the literature is very limited.

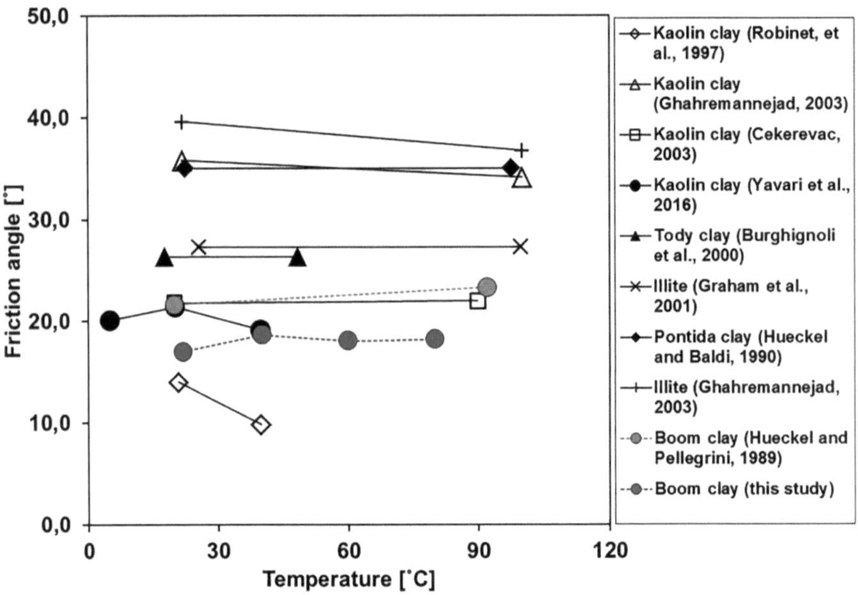

Fig. 2.25 Thermal effect on friction angle from literature and this study

2.3 THM Permeability Tests

Constant-head permeability tests were carried out in a temperature-controlled triaxial cell. The experimental study is carried out that intended to investigate (1) the anisotropic properties of hydraulic conductivity of Boom Clay, (2) the effect of heating–cooling cycles on the hydraulic conductivity and intrinsic permeability of Boom Clay, and (3) the effect of loading–unloading cycles on the hydraulic conductivity and intrinsic permeability of Boom Clay. The test condition is shown in Table 2.3.

2.3.1 Test Methodology and Procedures

2.3.1.1 Materials and Sample Preparation

The hydraulic conductivities of Boom Clay measured through various testing techniques exhibit similar values in the order of 10^{-12} m/s (Yu et al. 2013). To ensure a measurable flow in the constant-head method in the dense plastic clay, smaller samples with the standard diameter (38 mm) but reduced height (10 mm) were used (Fig. 2.26b). To account for the anisotropy of Boom Clay, samples were manually trimmed with parallel axes (horizontal sample) and perpendicular (vertical sample) to the bedding (Fig. 2.26a). The main geotechnical characteristics of samples are shown in Table 2.4.

Table 2.3 Test condition of THM-coupled permeability tests

Sample	Confining pressure (MPa)	Back pressure (MPa)	Temperature (°C)
Horizontal	2.5 → 5.5 → 2.5	1.0	23 → 40 → 60 → 80 → 60 → 40 → 23
Vertical	2.5 → 5.5 → 2.5	1.0	23 → 40 → 60 → 80 → 60 → 40 → 23

(a) Boom Clay core (b) sample

Fig. 2.26 Boom Clay samples for THM-coupled permeability tests. (**a**) Boom Clay core; (**b**) sample

Table 2.4 Initial condition for the specimens of THM-coupled permeability tests

Sample	Density (g/cm^3)	Water content (%)	Dry density (g/cm^3)	Void ratio	Saturation degree (%)
Horizontal	2.02	23.67	1.63	0.68	95.03
Vertical	2.04	23.88	1.64	0.67	97.37

As the previous THM tests conducted, SBCW was used for the re-saturation of the samples.

2.3.1.2 Test Procedures

The experimental procedure, for both the horizontal and vertical samples, involves eight stages after sample re-saturation (the re-saturation procedure see Sect. 1.2.1.4):

Stage 1: The sample was isostatically loaded to a confining pressure ($\sigma_1 = \sigma_2 = \sigma_3 = 2.5$ MPa, which is close to in situ effective stress).

Stage 2: A heating–cooling cycle (23 °C, the room temperature→40 → 60 → 80 → 60 → 40 → 23 °C) with a rate of 0.3 °C/h was applied.

Stage 3: The sample was isostatically loaded at a rate of 40 kPa/h from 2.5 MPa to 5.5 MPa (close to preconsolidation stress).

Stage 4: Another heating–cooling cycle similar to that in Stage 2 was applied.

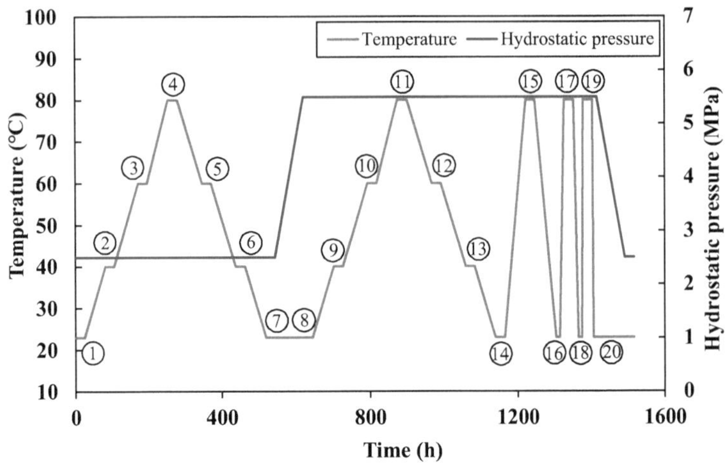

Fig. 2.27 Heating–cooling and loading–unloading procedures of constant-head test

Stages 5, 6, and 7: Three additional heating–cooling cycles (23 → 80 → 23 °C) with different rates (1, 5, and 20 °C/h, respectively) were applied.

Stage 8: The sample was isostatically unloaded back to 2.5 MPa at a rate of 40 kPa/h.

The detailed testing procedure is shown in Fig. 2.27. It would take more than 2 months to complete all eight stages for each sample. The stress states in which the constant-head tests are conducted are clearly marked in this figure (1–18). As long as the confining pressure or temperature reached the predetermined values in rows of Table 2.5, constant-head tests were carried out. Eighteen constant-head permeability measurements (the last column Table 2.5) were taken for both the horizontal and vertical samples. The back pressure was maintained at 1 MPa during the entire test at the bottom of the sample, while the top porous stone was put in contact with the atmospheric pressure by unscrewing the pipe connected to the top of the sample. The high back pressure was necessary to obtain satisfactory precision when measuring the flow rate and, hence, the hydraulic conductivity (Delage et al. 2000). The fluid injection for the test is SBCW.

2.3.2 *Test Results and Discussion*

2.3.2.1 Permeability Variation Under Heating–Cooling Cycles

The volume of water injected as a function of time is given in Fig. 2.28. Despite the reduced sample height and high gradient applied, it still required approximately 10 h to achieve a steady state of water flow. This is consistent with that described by Delage et al. (2000). Delage et al. (2000) found that 10 h is necessary to achieve permanent flow, and 15 h is needed to obtain a satisfactory determination of the

Table 2.5 THM path for the constant-head test

Experimental stage	Phase	Hydrostatic pressure (MPa)	Temperature (°C)	Constant-head testing sequences
1	Loading	2.5	23	①
2	Heating/cooling 1	2.5	23 → 40 → 60 → 80 → 60 → 40 → 23	① → ⑦
3	Loading	5.5	23	⑦ → ⑧
4	Heating/cooling 2	5.5	23 → 40 → 60 → 80 → 60 → 40 → 23	⑧ → ⑭
5	Heating/cooling 3	5.5	23 → 80 → 23	⑭ → ⑮ → ⑯
6	Heating/cooling 4			⑯ → ⑰ → ⑱
7	Heating/cooling 5			⑱ → ⑲ → ⑳
8	Unloading	2.5	23	⑳

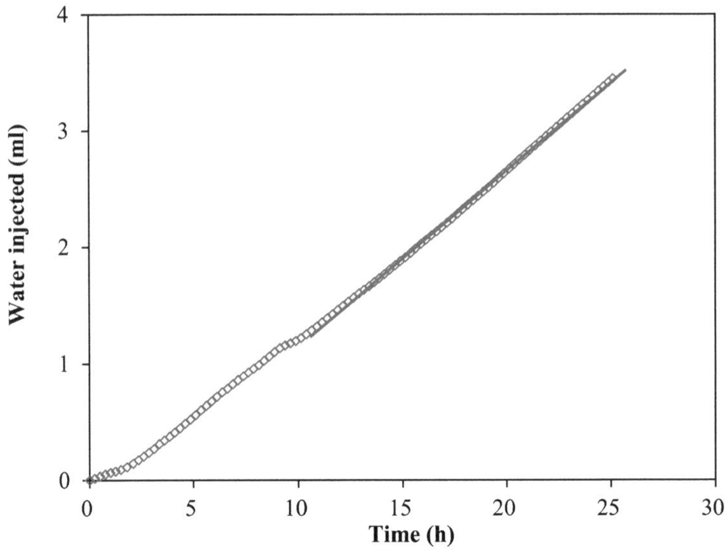

Fig. 2.28 Volume of water injected during a permeability test

slope of the curve that corresponds to a constant flow. The characteristics of the steady state of water flow through Boom Clay at different temperatures and confining pressures are presented in Fig. 2.29. The steady flow duration at each temperature level is approximately 10 h. A linear relationship can be obtained. The injected water volume within the same duration increases with increasing temperature, indicating a higher hydraulic conductivity at higher temperatures, while drops by one-third were observed when the confining pressures increased from 2.5 to 5.5 MPa. The hydraulic conductivity was calculated by applying Darcy's law. The test results were compared with the results of previous studies in Fig. 2.30. The same trend was observed between hydraulic conductivity and temperature in previous studies. Delage et al. (2000) measured the vertical hydraulic conductivity under a confining pressure of 2.5 MPa using the constant-head method. The authors found an increase in the hydraulic conductivity from 2.5×10^{-12} m/s to 6.2×10^{-12} m/s with temperature increasing from 20 °C to 90 °C. Similar results were reported by Chen et al. (2014), who measured the vertical hydraulic conductivity in a permeameter cell (non-loading system) and noted an increase in the hydraulic conductivity from 2.2×10^{-12} m/s to 7.4×10^{-12} m/s with an increase in temperature from 23 °C to 80 °C. Monfared et al. (2012) evaluated the permeability of Boom Clay (before and after compression) using a transient method. Unfortunately, the permeability before compression was only measured at room temperature (confining pressure of 3.25 MPa). There have been slight differences in the reported hydraulic conductivity values in different papers, which could be acceptable because of the different test methods and boundary conditions.

Figure 2.31 shows the variation of hydraulic conductivity of Boom Clay during two heating–cooling cycles. The figure shows that (1) at room temperature, the vertical (K_v) and horizontal (K_h) hydraulic conductivities are 1.73×10^{-12} m/s and

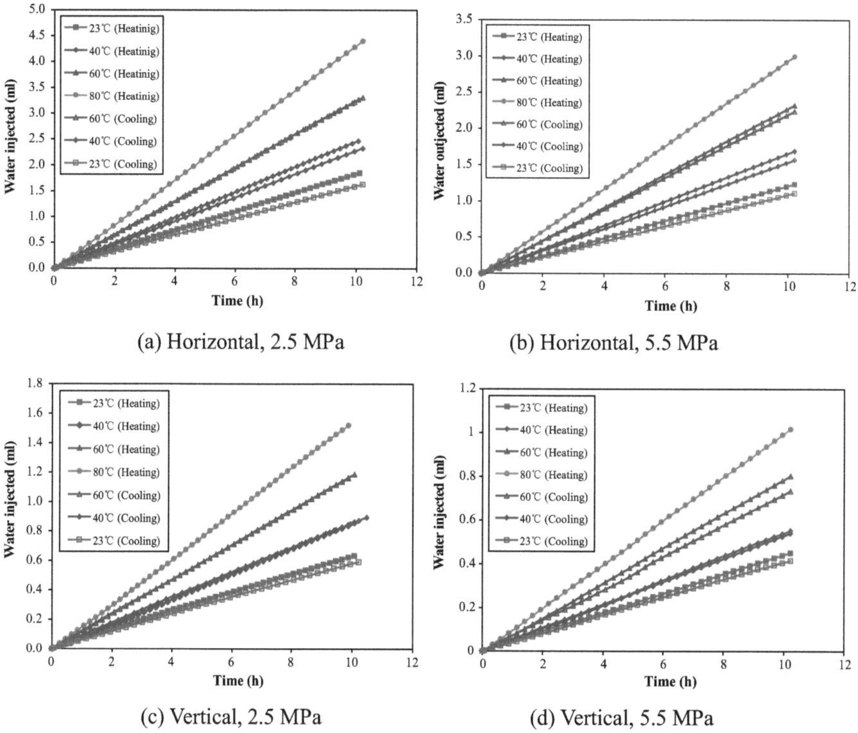

Fig. 2.29 Injected water flow-time relationships of steady-state experiments at different temperatures and stress. (**a**) Horizontal, 2.5 MPa; (**b**) horizontal, 5.5 MPa; (**c**) vertical, 2.5 MPa; (**d**) vertical, 5.5 MPa

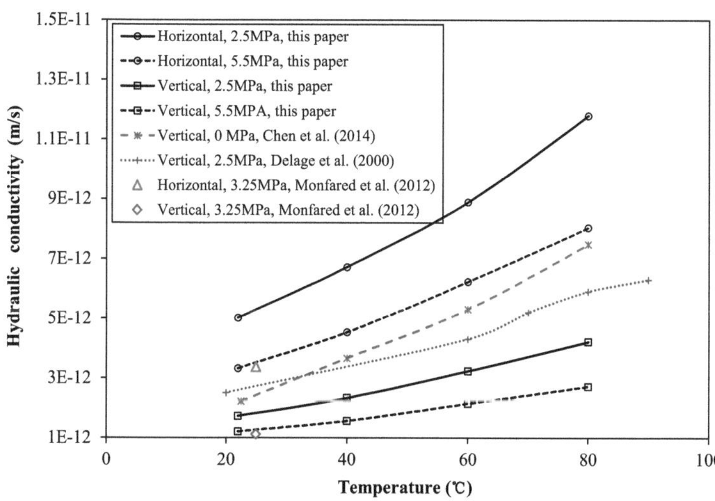

Fig. 2.30 Comparison of the hydraulic conductivity measured in this book and the results of previous studies (Delage et al. (2000); Monfared et al. (2012); Chen et al. (2014))

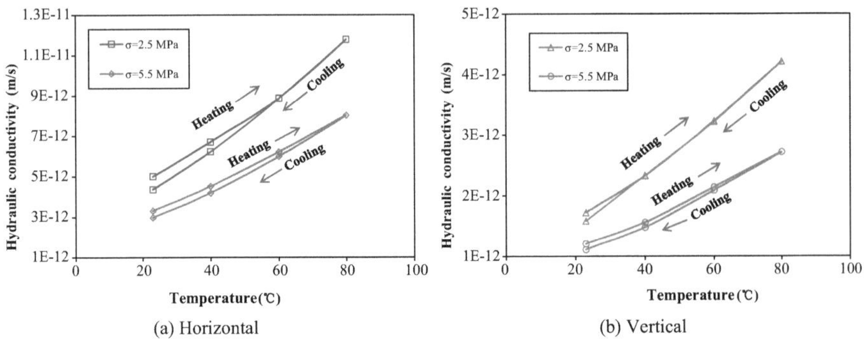

Fig. 2.31 Variation of the hydraulic conductivity during heating–cooling cycles. (**a**) Horizontal; (**b**) vertical

5.01×10^{-12} m/s, respectively, which falls within the range of measured in situ hydraulic conductivity ($1.7–2.39 \times 10^{-12}$ m/s for K_v and $4.1–5.2 \times 10^{-12}$ m/s for K_h (Bastiaens et al. 2006); (2) the anisotropic characteristic of hydraulic conductivity of Boom Clay concerning its bedding was confirmed. The horizontal hydraulic conductivity (parallel to bedding) is 2.8 times larger than the vertical hydraulic conductivity (perpendicular to bedding); (3) there is a positive and reversible relationship between hydraulic conductivity and temperature during the heating–cooling cycle, which is consistent with the study of Chen et al. (2014); (4) the hydraulic conductivity at 80 °C is about 2.4 times larger than the one at room temperature; and (5) the hydraulic conductivities along both horizontal and vertical directions drop by more than 30% when the confining pressure is increased from 2.5 to 5.5 MPa.

The intrinsic permeability, k', was computed according to:

$$k' = \frac{K\mu_w}{\gamma_w} \tag{2.2}$$

where K is the hydraulic conductivity, μ_w is the water viscosity, and γ_w is the unit weight of water. The thermal variation of γ_w and μ_w of "pure water" is available in Delage et al. (2000) and Cho et al. (1999), respectively. The variations of the intrinsic permeability of Boom Clay under different temperatures and confining pressures are shown in Fig. 2.32. It can be seen that during the heating–cooling cycle, the intrinsic permeability decreases by approximately 10% (for all tests) during the heating phase and later increases slightly during the cooling phase. This may be attributed to the thermal volume change behavior of Boom Clay. The plastic thermal contraction of samples during the heating phase results in a decrease in the intrinsic permeability of Boom Clay.

2.3.2.2 Permeability Variation under Loading–Unloading Cycles

The variations of the hydraulic conductivity and intrinsic permeability at room temperature during the loading–unloading cycle are shown in Figs. 2.33 and 2.34. The results show that during the loading–unloading cycle, there is a negative and

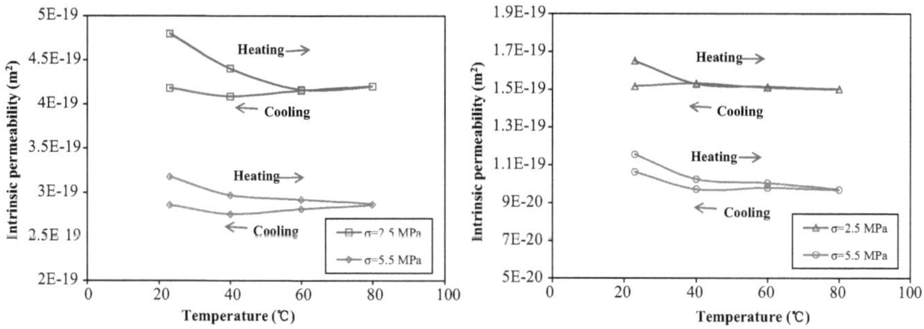

(a) Horizontal (b) Vertical

Fig. 2.32 Variation of the intrinsic permeability during heating–cooling cycles. (**a**) Horizontal; (**b**) vertical

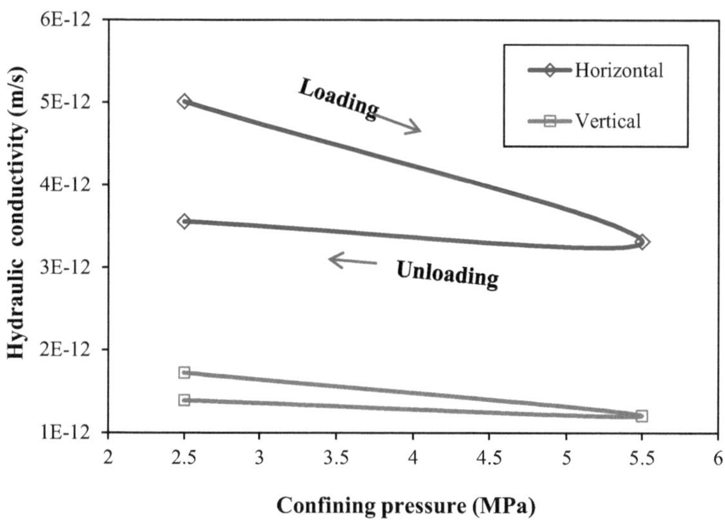

Fig. 2.33 Variation of the hydraulic conductivity at room temperature during loading–unloading cycle

irreversible relationship between hydraulic conductivity and hydrostatic pressure. Since the hydraulic conductivity is reversible during the heating–cooling phase (under both 2.5 and 5.5 MPa), the irreversible variation during the loading–unloading phase can be attributed to the mechanically induced plastic volume change behavior.

The different effects of the heating–cooling cycle and the loading–unloading cycle on the variation of intrinsic permeability can be observed in Figs. 2.32 and 2.34. For instance, at a hydrostatic pressure of 2.5 MPa, the horizontal intrinsic permeability changed from 4.80×10^{-19} m^2 to 4.20×10^{-19} m^2 upon heating (23–80 °C), and then slightly changed to 4.18×10^{-19} m^2 upon cooling to room temperature. This means that the intrinsic permeability varies by approximately

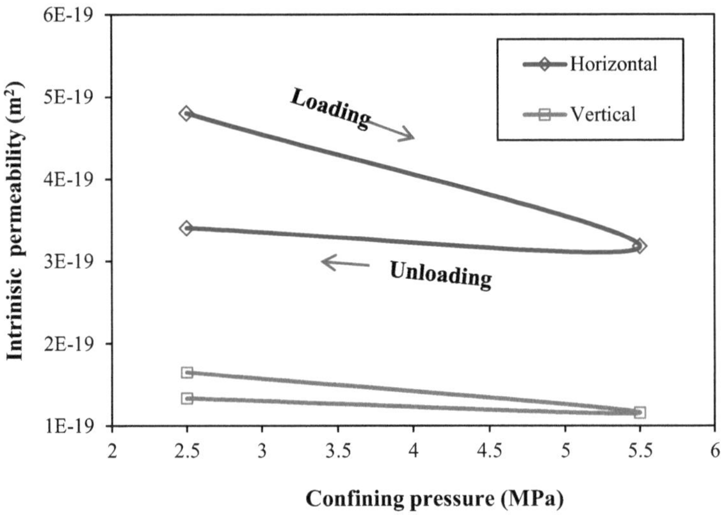

Fig. 2.34 Variation of the intrinsic permeability at room temperature during loading–unloading cycle

12% during the heating–cooling cycle within the temperature range of 23 °C and 80 °C. By contrast, during the loading–unloading cycle within the range of 2.5–5.5 MPa, the horizontal intrinsic permeability varies by more than 30%. This indicates that the thermally induced volume change within a reasonable temperature variation interval of a future repository (room temperature to 80 °C) is much smaller compared to the mechanically induced volume change. It is important to note that the stress levels used in the tests are close to the in situ effective stress (2.5 MPa) and preconsolidation stress (5.5 MPa). Hence, from the perspective of radioactive waste disposal engineering for the cases considered here, the thermally induced intrinsic permeability change of Boom Clay is much smaller compared to the mechanically induced intrinsic permeability change.

2.3.2.3 Effect of Heating Rate on Hydraulic Conductivity

The effect of heating rate on hydro-mechanical parameters of Boom Clay is very important, especially for the determination of heating rate of in situ heating tests. Sultan (1997) suggested that the shape of the contraction curve was affected by the cooling rate. Test results reported by Cui et al. (2000) also show that the cooling rate has a significant effect on the slope of the temperature (T)—thermally induced volumetric strain (ε_{vT}) diagram.

Figure 2.35 shows the variation of the hydraulic conductivity at 80 °C reached at different heating rates obtained by this study (stages 5–7). The results show that the hydraulic conductivity stays constant with different heating rates, suggesting that the thermally induced volume change behavior has an insignificant effect on the

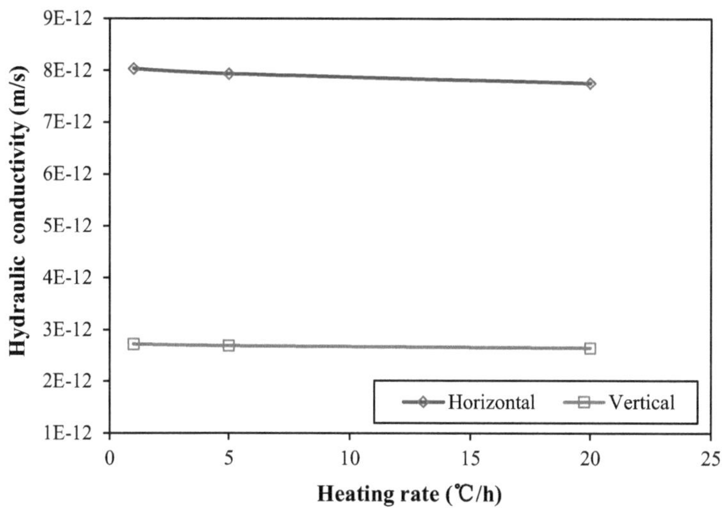

Fig. 2.35 Variation of the hydraulic conductivity at 80 °C with different heating rates

hydraulic conductivity of Boom Clay. The change in the hydraulic conductivity of Boom Clay with temperature may be mostly affected by changes in the viscosity and density of the pore fluid with temperature. Further investigations into this aspect are necessary, and a difference between the test results under drained and undrained conditions is expected.

2.4 Thermal-Induced Microstructure Change

2.4.1 Test Methodology

Nuclear magnetic resonance (NMR) tests provide information about the water content and pore distribution of soil. The NMR analyzer, as illustrated in Fig. 2.36, is mainly composed of a permanent magnet, sample tube, RF (radio frequency) system, and the data collecting and analyzing system. The strength of the permanent magnet is 0.52 T and the magnet temperature is maintained at 32 ± 0.01 °C in order to ensure the uniformity and stability of the main magnetic field. Its design, operation, and service condition were discussed in detail by Tian and Wei (2014).

The dimensions of the samples in the NMR analyzer are the same as the THM-coupled permeability tests. Sample re-saturation has been done under in situ effective stress, and a satisfactory value of the Skempton's B coefficient was obtained. Re-saturation samples were surrounded by preservative film and heated at 23, 40, 60, and 80 °C. The equilibrium time of each temperature level is about 2 h. The specimen was then moved rapidly to the sample tube for NMR experiments. Bayer et al. (2010) summarized the applications of the NMR technique in soil science. The

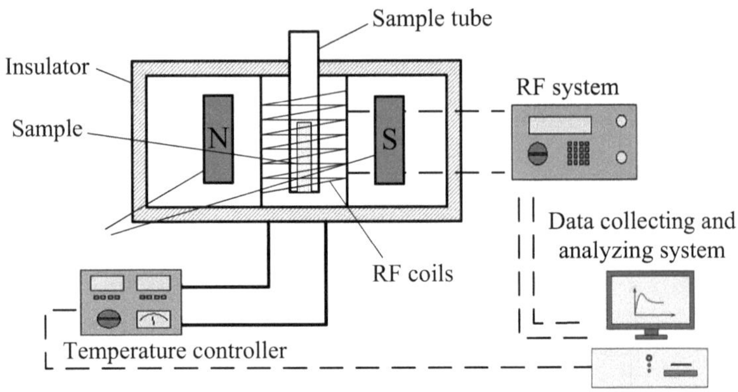

Fig. 2.36 Schematic diagram of the nuclear magnetic resonance analyzer

measured signal of the NMR test is called the free induction decay (FID). The FID curve was determined by using a CPMG (Carr-Purcell-Meiboom-Gill) sequence in the present study. The FID curve contains rich information about the water content and water-occupied pore-size distribution of the porous medium. The initial FID peak value is directly proportional to the number of water protons in the sample (Ishizaki et al. 1996). Hence, the initial FID peak measurements can determine the water content of the sample. Moreover, the shape of the FID curve is associated with the transverse relaxation time (T_2) of pore water protons. The T_2 relaxation time distribution curve can be obtained by the inverse Laplace transform. The peak area (dimensionless) under T_2 distribution curve represents the water content of the corresponding T_2 range (Bird et al. 2005). The transverse relaxation T_2 of the water protons is proportional to the water-filled pore size, which means water in small pores relaxes faster than that in large pores:

$$\frac{1}{T_2} \approx \rho_2 \frac{\alpha''}{R} \tag{2.3}$$

where R is the pore radius, ρ_2 is the transverse relaxivity, and α'' is the shape factor (1, 2, 3 for planar, cylindrical, and spherical pore geometry) (Godefroy et al. 2001).

2.4.2 *Thermal-Induced Microstructure Changes Behavior of Boom Clay*

Figure 2.37 shows the temperature effect on the T_2 relaxation time distribution of saturated Boom Clay. The unchanged peak area of four curves demonstrates that the sample stays saturated during the whole test (Bird et al. 2005). Moreover, the T_2 relaxation time distribution curve moves to the right with increasing temperature,

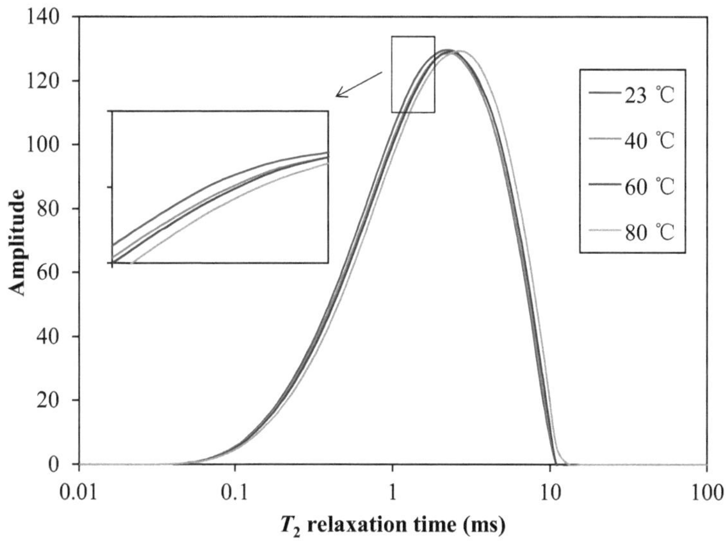

Fig. 2.37 Transverse relaxation time distribution curve of water in saturation Boom Clay

indicating an increase in the T_2 relaxation time. The increasing T_2 relaxation time signifies that the water-occupied pore size in saturated Boom Clay becomes larger with increasing temperature. These results are similar to the previous results from X-ray diffraction techniques (Pons et al. 1994; Thomas et al. 1994) and Electron Microscopic Examination (Pusch and Güven 1990), which showed that heating produced larger voids among clay particles. It is worth noting that the above-mentioned phenomena occurred when clays were heated under unstressed conditions. The larger internal particle spaces produced by heating consequently would lead to microstructural weakening and a decrease in stiffness and strength.

2.5 Mechanisms of Thermal Effects on Hydro-Mechanical Properties

Heat treatment of clays would change several material properties, such as hydraulic conductivity, friction angle, and cohesion. But what makes the material properties of some clays sensitive to temperature is still unknown.

In most studies, the increase in hydraulic conductivity with temperature has been attributed solely to the decrease in viscosity of fluid with temperature (Delage et al. 2000; Cho et al. 1999). Therefore, the hydraulic conductivity of Boom Clay was predicted by considering the variation of water viscosity and density, taking the hydraulic conductivity experimentally measured at room temperature as a starting point.

$$K(T) = \frac{\mu_w(T_0)\rho_w(T_0)}{\mu_w(T)\rho_w(T)} K(T_0) \tag{2.4}$$

in which

$$\mu_\mathrm{w} = 2.9\times10^{-11}T^4 - 9.17\times10^{-9}T^3 + 10^{-6}T^2 - 5.56\times10^{-11}T + 1.7862\times10^{-3} \tag{2.5}$$

$$\rho_\mathrm{w} = -5.5\times10^{-6}T^2 + 2.28\times10^{-5}T + 0.99997 \tag{2.6}$$

Figure 2.38 shows the comparison between the hydraulic conductivity determined experimentally and those predicted based on water viscosity and density changes. The figure confirms that the changes in water properties are the main driving factors in the variation of hydraulic conductivity of Boom Clay with temperature.

However, it is worth noting that the increase in hydraulic conductivity with temperature obtained experimentally is slightly lower than that predicted based on water viscosity and density changes. The discrepancy between the hydraulic

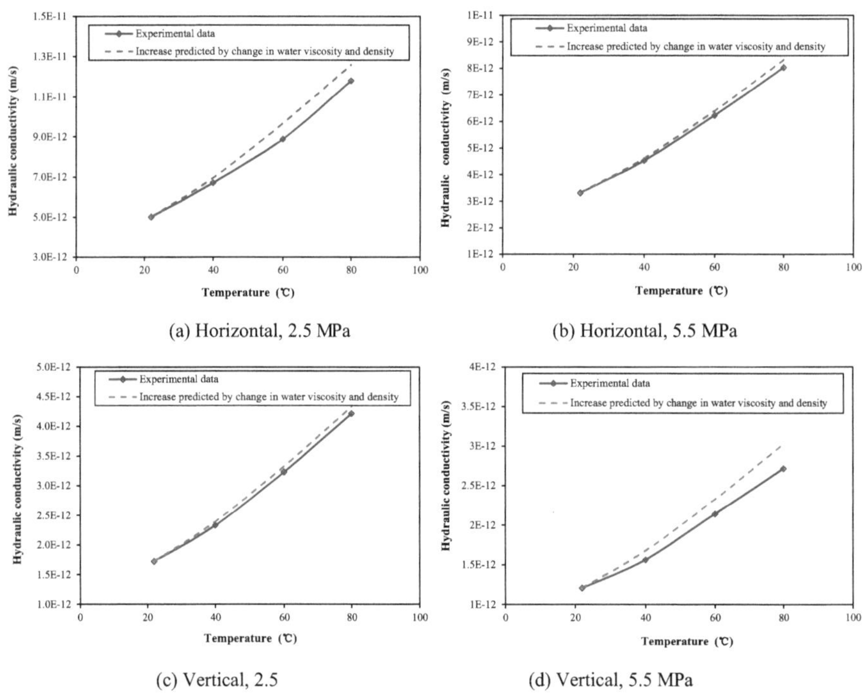

Fig. 2.38 Comparison of the hydraulic conductivity experimentally determined and predicted, only considering the variations of water viscosity and density with temperature changes. (**a**) Horizontal, 2.5 MPa; (**b**) horizontal, 5.5 MPa; (**c**) vertical, 2.5 MPa; (**d**) vertical, 5.5 MPa

conductivity experimentally measured and predicted may be attributed to the changes in clay porosity with temperature. The results of NMR tests found that the larger internal particle spaces produced by heating consequently would lead to microstructural weakening and the decrease in stiffness. The samples would be compacted when heated under hydrostatic stress during the THM-coupled permeability tests. Therefore, the decrease in the intrinsic permeability (Fig. 2.32) is due to the decrease in void ratio with temperature. Furthermore, it was noted that the thermal volume change behavior also has an effect on the hydraulic conductivity of Boom Clay, along with the water viscosity and density.

The temperature effects on the strength of clay have always been a subject of discussion. Many researchers explained such discussion from the modification of cohesion, irreversible volume changes behavior (Cui et al. 2009), depending on the overconsolidation ratio (Kuntiwattanakul et al. 1995) and thermo-mechanical loading paths (Noble 1968). Robinet et al. (1996) speculated that a high temperature-induced irreversible microstructural changes in clay, leaving unstable interactions between clay particles and thus decreasing its strength. The speculation can be proved by the NMR tests in this book. Therefore, it is logical to conclude that the strength of clay is affected comprehensively by two aspects: thermal softening and thermal hardening. When the thermal softening effect related to the microstructural weakening cannot be balanced by the volume contraction, the increase in temperature would decrease the clay strength. Conversely, the large thermal volume contraction makes the thermal hardening effect dominant. The clay strength increases with the increasing temperature.

However, more research is required to better understand the mechanism of thermal effects.

2.6 Conclusions

A series of laboratory tests were performed on the Boom Clay to find the nature of the thermo-hydro-mechanical response of this material. All the tests were realized on saturated samples and under various stress and thermal conditions.

1. The THM-coupled undrained triaxial compression tests show that the overconsolidation ratio (OCR) has a significant effect on the contracting/expanding nature of the volume change of a sample heated under a constant load. A decrease in shear strength of Boom Clay with temperatures has been found. This weakening in shear strength was attributed to the reduction of cohesion and elastic modulus. However, thermal effects on the friction angle of Boom Clay are unclear. Pore pressure buildup upon heating under undrained conditions was found to be dependent on the initial mean effective stress.

2. The THM-coupled permeability tests were carried out to investigate the variation of hydraulic conductivity of Boom Clay during heating–cooling cycles and loading–unloading cycles. On the one hand, the anisotropic characteristic of hydraulic conductivity of Boom Clay with respect to its bedding was confirmed. On the other hand, there is a positive and reversible relationship between the hydraulic conductivity and temperature and a negative and irreversible relationship between the hydraulic conductivity and hydrostatic pressure.
3. Furthermore, nuclear magnetic resonance (NMR) tests were conducted on heated Boom Clay samples to investigate thermal effects on the microstructural properties of Boom Clay. The results of NMR tests found that larger interparticle spaces would be produced by heating and would then lead to microstructural weakening of Boom Clay.

Chapter 3
A Thermo-Hydro-Mechanical-Damage Model for Boom Clay

3.1 Introduction

Experimental studies in Chap. 2 show that temperature has a significant influence on the hydro-mechanical behavior of Boom Clay. In the past decade, our research team has tried to establish thermo-elasto-plastic damage models for Boom Clay (Jia 2009; Jia et al. 2011; Yu 2010; Gong 2015). The advantage of the theoretical models we have established is that most of the models can predict the major aspects of thermo-mechanical behavior, which were validated by simulating the field measurements from the laboratory and the HADES underground research laboratory (URL). The preliminary thermo-elasto-plastic damage model (Gong 2015) did not consider the anisotropic properties of Boom Clay and could not provide good predictions of the four stages of typical stress–strain behavior of Boom Clay, especially the plastic flow stage, and it cannot reproduce the evolutions of excess pore pressure during the THM triaxial compression tests. In this chapter, we developed a transversely isotropic thermo-hydro-mechanical-damage (THMD) model, with the consideration of the elastic damage evolution, plastic damage evolution, thermal damage evolution, plastic hardening law, and hydraulic conductivity evolution in the framework of the Drucker–Prager Cap model. The new THMD constitutive model has been implemented in the ABAQUS code through the subroutine USDFLD. The calibration of parameters has been done using the test data obtained in Chap. 2.

W. Chen et al., *Thermo-Hydro-Mechanical Coupling in Clay Rock*,
https://doi.org/10.1007/978-981-95-2562-1_3

3.2 Stress–Strain Behavior of Boom Clay Under Different Temperatures

Typical stress–strain behavior of Boom Clay under different temperatures is schematized in Chap. 2 and shown in Fig. 3.1. It can be broken down into four stages: elastic stage (OA), strain-hardening stage (AB), strain-softening stage (BC), and plastic flow stage (CD).

A piecewise approach is adopted in order to fully describe the mechanical behavior of Boom Clay. Specifically, three parts, including elastic damage evolution law, plastic hardening law, and plastic damage evolution law, are introduced in the model. Moreover, the thermal damage variable is used to describe the stiffness and strength degradation behavior due to temperature increase.

3.3 Criterion of Yielding

Baldi et al. (1991) summarized a great deal of experimental data on Boom Clay, and its yield surface is shown in Fig. 3.2. It is obvious that the yield surface of the Modified Cam-Clay model substantially overestimates the strength of heavily overconsolidated clay. Baldi et al. (1991) suggested that the Drucker–Prager Cap model can describe the plastic mechanical behavior of Boom Clay well. The Drucker–Prager Cap model is widely applied in rock and soil mechanics to account for shear failure and consolidation behavior of soft rocks and clays. One of the reasons for the

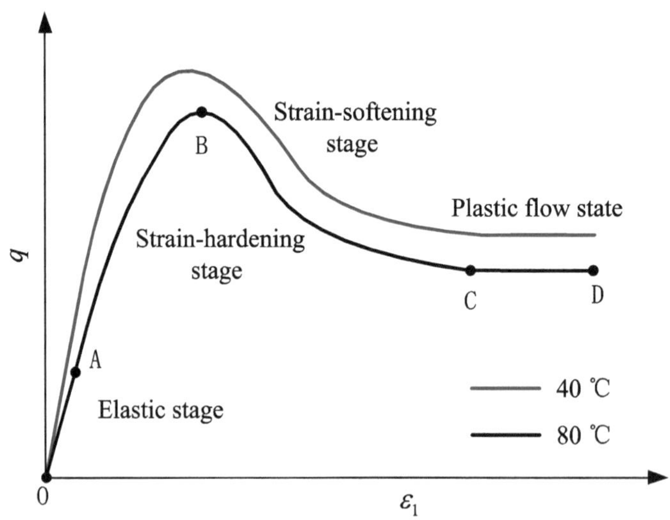

Fig. 3.1 Stress–strain curve of Boom Clay

Fig. 3.2 The shape of the yield surfaces for Boom Clay

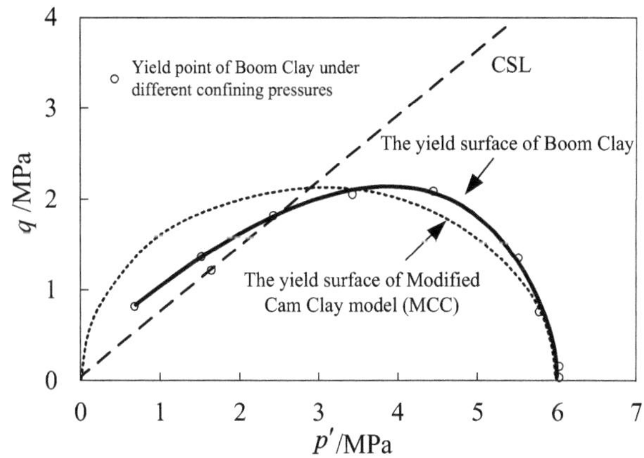

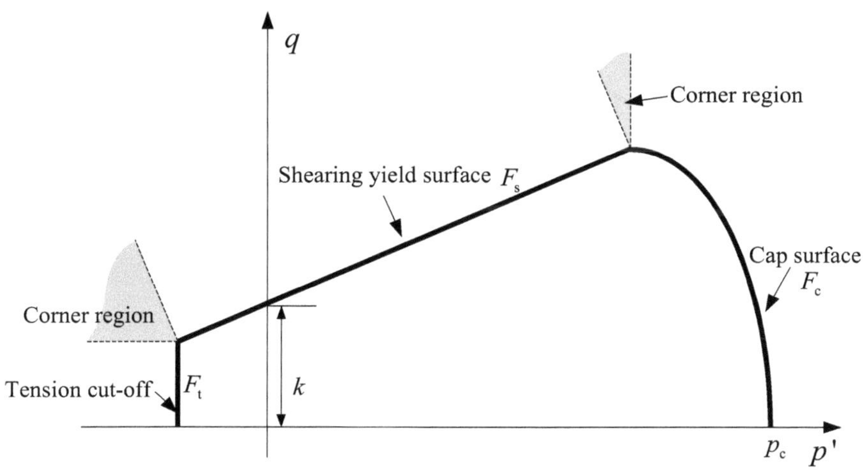

Fig. 3.3 Typical cap model with corner regions in the p–q plane (Dolarevic and Ibrahimbegovic 2007)

wide application is that it is described by a simple mathematical expression and thus is easy to understand and simple to apply.

The typical yield surface for the cap model is shown in Fig. 3.3. It consists of three portions: a fixed Drucker–Prager failure envelope, an elliptic cap surface, and a tension cut-off plane. The yield function on intersection points of these surfaces is not smooth, and the corner regions can cause numerical singularity (Dolarevic and Ibrahimbegovic 2007).

The modified Drucker–Prager Cap embedded in the ABAQUS code has solved this problem by introducing elliptical surfaces on both the shearing and the compression side of the potential surface (Abaqus 2006). The potential surface is second-order continuous, and a stable tangent operator can be derived, which ensures a robust numerical implementation.

The yield surfaces of Drucker–Prager Cap are combined with the two parts, as shown in Fig. 3.4. The shearing yield surface is the same as the Drucker–Prager yield surface and an elliptical cap surface. The expression of the shearing yield surface is:

$$F_s = q - \alpha p' + k = 0 \tag{3.1}$$

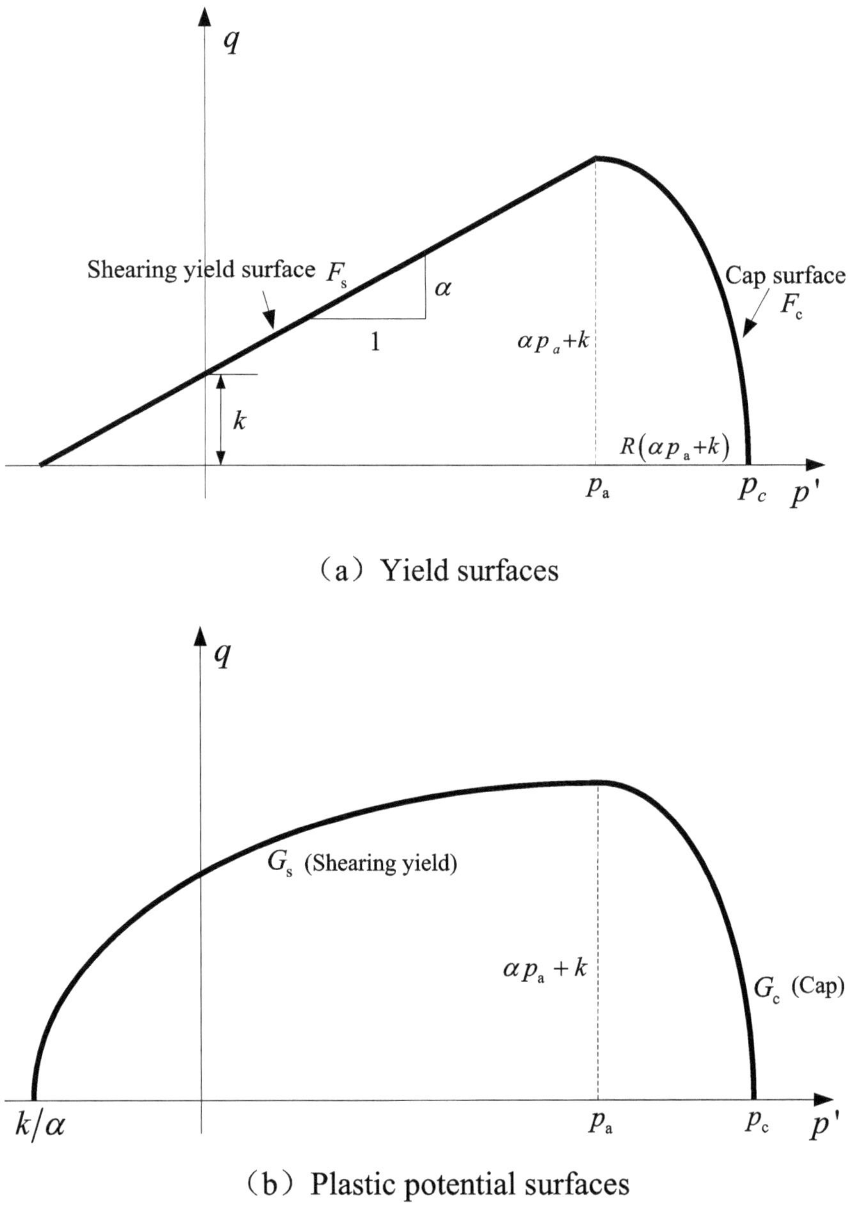

Fig. 3.4 The yield surfaces and plastic potential surface of the Drucker–Prager Cap model. (**a**) Yield surfaces; (**b**) plastic potential surfaces

where α and k are, respectively, the slope and intercept of the shearing surface in meridional plane, p' is the mean stress, and q is the deviatoric stress.

$$p' = \frac{1}{3}\mathrm{tr}(\boldsymbol{\sigma}) = \frac{1}{3}(\sigma_x + \sigma_y + \sigma_z) \tag{3.2}$$

$$q = \left(\frac{3}{2}\mathbf{s}:\mathbf{s}\right)^{\frac{1}{2}} = \left(\frac{3}{2}\left(s_x^2 + s_y^2 + s_z^2\right) + 3\tau_{xy}^2 + 3\tau_{xz}^2 + 3\tau_{yz}^2\right)^{\frac{1}{2}} \tag{3.3}$$

$$\mathbf{s} = \boldsymbol{\sigma} - \frac{1}{3}p\boldsymbol{\delta} \tag{3.4}$$

where $\mathbf{s}$ is the deviatoric stress tensor. α and k can be expressed as:

$$\alpha = \frac{6\sin\varphi}{(3 - \sin\varphi)} \tag{3.5}$$

$$k = \frac{6c}{(3 - \sin\varphi)} \tag{3.6}$$

where φ, c are the friction angle and cohesion, respectively.

The expression of the elliptical cap yield surface is:

$$F_c = \sqrt{q^2 + \left(\frac{p' - p_a}{R}\right)^2} - (\alpha p_a + k) = 0 \tag{3.7}$$

where R is a parameter representing the shape of the cap, and p_a is given by

$$p_a = \frac{p_c - Rk}{1 + \alpha R} \tag{3.8}$$

where p_c is the preconsolidation pressure.

The plastic potential surface in the meridional plane is made up of two elliptical portions. The plastic potential in shearing failure is given by

$$G_s = \sqrt{\left[\alpha(p' - p_a)\right]^2 + q^2} = 0 \tag{3.9}$$

The plastic potential in the shearing cap region is given by

$$G_c = \sqrt{(p' - p_a)^2 + (Rq)^2} = 0 \tag{3.10}$$

3.3.1 Elastic Damage and Thermal Damage Evolution Law

Damage mechanics are widely used in theoretical models to describe the behavior of various materials. The theory is supported by the physical idea that crack initiation is preceded by a progressive internal deterioration of the material (Mahnken

2002; Litewka and Debinski 2003), which induces a loss of strength in terms of strain as well as in terms of remaining life (Chaboche 1981). It provides a method to understand the weakening mechanism at the micromechanical scale due to nucleation and coalescence of voids, cavities, and cracks. Generally, damage variables are introduced to reflect the material weakening in an element at a macro-mechanical scale of the classical theory of continuum mechanics (Gong 2015).

Damage due to the initiation and growth of microcracks has also been observed in the Boom Clay (Bastiaens et al. 2006). However, limited theoretical models were proposed to describe the deterioration of Boom Clay, though certain laboratory tests and in situ measurements on Boom Clay have proved the existence of damage.

3.3.2 Elastic Damage Evolution Law

The stress–strain relation of Boom Clay has obvious nonlinear properties before yielding. After unloading during the "elastic stage", the strain does not restore completely, which shows that damage has already occurred in the so-called elastic stage (Giraud and Rousset 1996). The evolution of the elastic moduli with damage is shown in Fig. 3.5.

It is assumed that the elastic damage evolution is temperature independent. Yu et al. (2014) chose the energy factor $\overline{e}$ as the driving force of elastic damage evolution and the expression of the elastic damage criterion is

$$D_{\mathrm{e}} = \beta_{\mathrm{M}} \left(\overline{e} - \overline{e}_{\mathrm{oe}} \right) \tag{3.11}$$

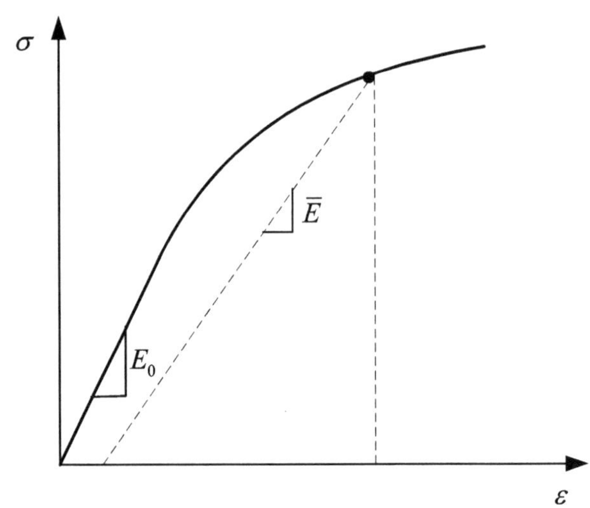

Fig. 3.5 Evolution of elastic modulus

in which

$$\overline{e} = \sqrt{\varepsilon : \boldsymbol{C} : \varepsilon} \tag{3.12}$$

where β_{M} is a damage parameter, $\overline{e}$ is the energy factor, and $\overline{e}_{\mathrm{oe}}$ is the energy factor threshold for the onset of elastic damage. After the elastic damage is activated, $\overline{e}_{\mathrm{oe}}$ presents the maximum $\overline{e}$ in the history. The evolution of the elastic damage variable is shown in Fig. 3.6.

3.3.3 Thermal Damage Evolution Law

The thermal damage is considered a weakening effect on the mechanical behavior of clay due to temperature. The model assumes that the mechanical damage evolution and the thermal damage evolution are mutually independent. Gong (2015) suggested that the expression of the thermal damage is

$$D_{\mathrm{T}} = 1 - \exp\left[-\beta_T \left(T - T_0\right)\right] \tag{3.13}$$

in which

$$T = \begin{cases} T, & T > T_{\mathrm{c}} \\ T_{\mathrm{c}}, & T \le T_{\mathrm{c}} \end{cases} \tag{3.14}$$

where T is the temperature in °C, T_0 is the initial temperature, T_{c} is the highest temperature in the history, and β_T is a constant.

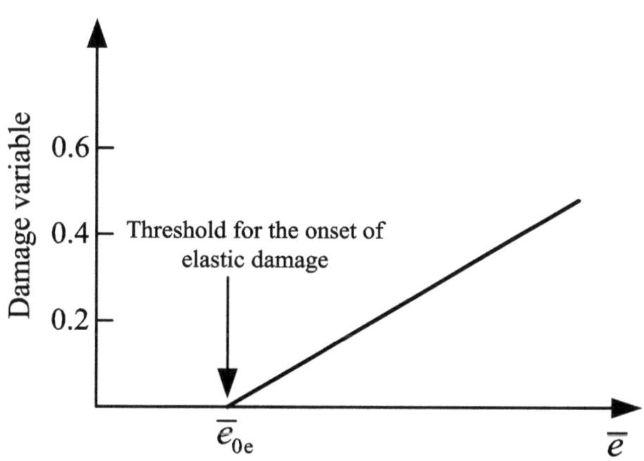

Fig. 3.6 Evolution of the elastic damage variable

Lots of data indicate that Boom Clay presents transversely isotropic mechanical properties. The elastic matrix of transversely isotropic materials can be expressed as:

$$[\mathbf{C}_0]^{-1} = \begin{bmatrix} \frac{1}{E_{v0}} & \frac{-\nu_{vh}}{E_{v0}} & \frac{-\nu_{vh}}{E_{v0}} & 0 & 0 & 0 \\ \frac{-\nu_{vh}}{E_{v0}} & \frac{1}{E_{h0}} & \frac{-\nu_{hh}}{E_{h0}} & 0 & 0 & 0 \\ \frac{-\nu_{vh}}{E_{v0}} & \frac{-\nu_{hh}}{E_{h0}} & \frac{1}{E_{h0}} & 0 & 0 & 0 \\ 0 & 0 & 0 & \frac{1}{G_{vh0}} & 0 & 0 \\ 0 & 0 & 0 & 0 & \frac{2(1+\nu_{hh})}{E_{h0}} & 0 \\ 0 & 0 & 0 & 0 & 0 & \frac{1}{G_{vh0}} \end{bmatrix} \tag{3.15}$$

in which E_{h0}, E_{v0} are the initial vertical and horizontal elastic modulus, respectively; ν_{hh} and ν_{vh} are Poisson ratio in the horizontal plane and the vertical plane, respectively; and G_{vh0} is the initial shear modulus in the vertical plane.

The elastic modulus evolution with elastic damage variable at various temperatures can be written as

$$\mathbf{C} = (1 - D_e)(1 - D_T)\mathbf{C}_0 \tag{3.16}$$

3.4 Plastic Hardening Law and Plastic Damage Evolution

3.4.1 Plastic Hardening Law

It is generally accepted that plastic damage occurs in the whole plastic stage. However, the stress–strain curve (Fig. 3.1) indicates that, at a macro-scale, the mechanical behavior during the whole plastic stage (stage ABCD) is quite different. Stage AB represents hardening plastic behavior, while stage BCD shows softening behavior. It is very complicated to describe the hardening behavior only by plastic damage. Therefore, a suitable plastic hardening law is first introduced to describe the plastic hardening behavior.

Based on the laboratory data of triaxial tests in Chap. 2, the following plastic hardening parameter following a nonlinear hyperbolic rule is proposed (Gong 2015):

$$h_p = h_{p0} + \frac{b_p(1 - h_{p0})\xi_p}{(1 - h_{p0}) + b_p\xi_p} \tag{3.17}$$

where h_{p0} and b_p are the constants, and ξ_p is the equivalent deviatoric plastic strain and is given by

$$\begin{cases} \dot{\xi}_p = \sqrt{\frac{2}{3}\dot{\mathbf{e}}_p : \dot{\mathbf{e}}_p} \\ \mathbf{e}_p = \varepsilon_p - \frac{1}{3}\mathrm{tr}(\varepsilon_p)\delta \\ \xi_p = \xi_p\big|_0 + \int_0^t \dot{\xi}_p dt \end{cases} \tag{3.18}$$

where $\mathbf{e}_p$ is the plastic strain.

3.4.2 Plastic Damage Evolution

For the softening and plastic flow stage (stage BCD in Fig. 3.1), plastic damage is significant. A plastic damage variable D_p is introduced into the instantaneous plastic hardening law to describe the shrinkage of the yield surface caused by the accumulation of plastic strain. The evolution of D_p takes the following form:

$$D_p = \bar{D}_p - \bar{D}_p \exp\left(-\beta_p \xi_p^a\right) \tag{3.19}$$

in which $\bar{D}_p$ is the upper limit value of D_p, and a and β_p are the model parameters. The relationship between a, β_p, and D_p is shown in Fig. 3.7.

Based on the results of THM-coupled triaxial tests, the temperature effects on the friction angle of Boom Clay are not significant. However, the cohesion has a certain decrease as the temperature increases from 22 to 80 °C. Schmertmann et al. (1960), Lambe et al. (1960), and Hajiabdolmajid et al. (2002) have also claimed that the friction angle of geomaterials is not decreased in the triaxial compression tests, and that the softening phenomenon is caused by the reduction of the cohesion (Fig. 3.8). Based on a careful review of previous studies, the model assumes that:

(1) The cohesion is controlled by the hardening variable, the plastic damage variable, and the thermal softening factor. It exhibits a transition from hardening to softening with the increase of axial strain:

$$c = (1 - D_T)(1 - D_p) h_p c_0 \tag{3.20}$$

where c_0 is the initial value of cohesion.

(2) The friction angle is unaffected by the thermal damage factor and follows nonlinear hyperbolic hardening's rule.

$$\varphi = \arctan\left(h_p \tan\varphi_0\right) \tag{3.21}$$

where φ_0 is the initial value of friction angle.

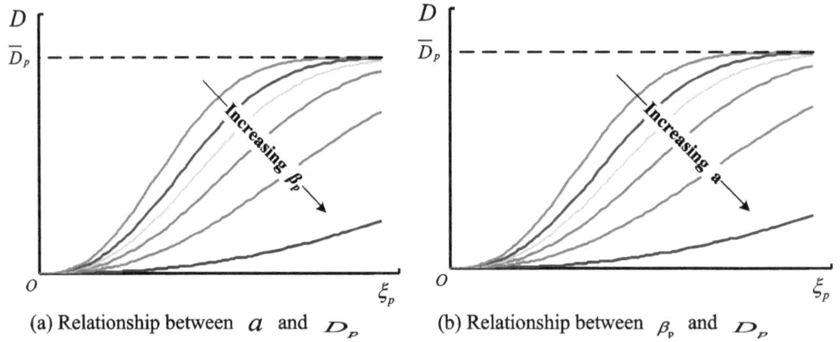

Fig. 3.7 Relationship between model parameters and plastic damage variable. (a) Relationship between a and D_p; (b) relationship between β_p and D_p

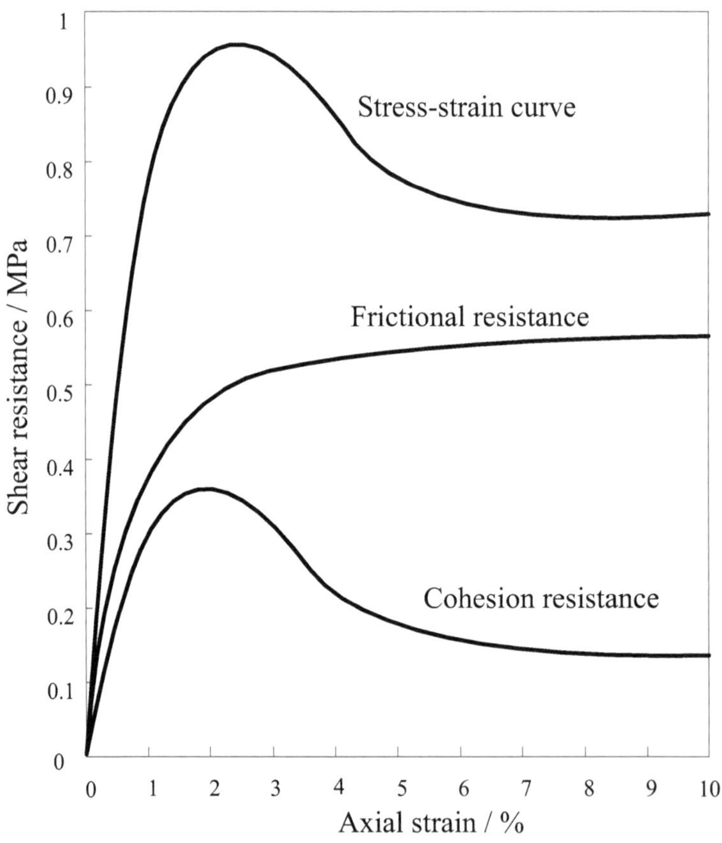

Fig. 3.8 Evolution of cohesion and friction angle of geomaterials during triaxial test (Schmertmann and Osterberg 1960)

(3) The preconsolidation pressure is comprehensively affected by the temperature and the irreversible volumetric strain

$$p_{\mathrm{c}} = (1 - D_{\mathrm{T}}) \exp\left(c_{\mathrm{p}} \varepsilon_{\mathrm{p}}^{\mathrm{v}}\right) p_{\mathrm{c0}} \tag{3.22}$$

where p_{c0} is the initial value of preconsolidation pressure, c_{p} is constant, and $\varepsilon_{\mathrm{p}}^{\mathrm{v}}$ is the plastic volumetric strain.

$$\varepsilon_{\mathrm{p}}^{\mathrm{v}} = \frac{1}{3} \mathrm{tr}\left(\boldsymbol{\varepsilon}_{\mathrm{p}}\right) \tag{3.23}$$

where $\boldsymbol{\varepsilon}_{\mathrm{p}}$ is the plastic strain tensor.

The yield surface of the model in the p'–q–T planes can be expressed by Fig. 3.9

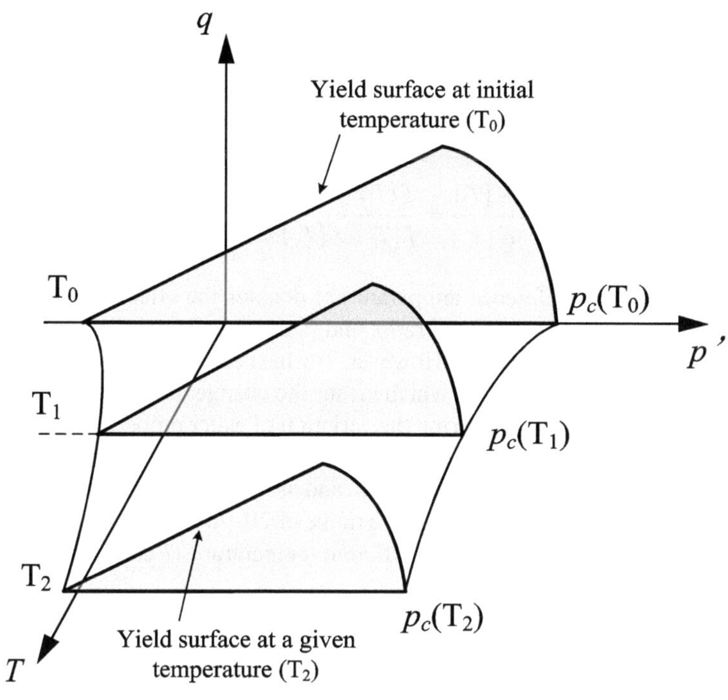

Fig. 3.9 Yield surface in p'–q–T planes

3.5 Hydraulic Conductivity Evolution Model

3.5.1 *Temperature Effect on the Hydraulic Conductivity of Boom Clay*

Under isothermal conditions, a widely used relationship between the hydraulic conductivity and physical properties of the pore fluid and clay mass for saturated clay is referred to as the Kozeny–Carman equation (Carman 1937).

$$K = C\left(\frac{g}{\mu_w \rho_w}\right)\frac{n^3}{(1-n)^2 S^2 G_s^{\,2}} \tag{3.24}$$

where C is an empirical parameter that is influenced by the tortuosity and shape of the flow channels, g is the gravitational constant, μ_w is the dynamic viscosity of water, ρ_w is the mass density of water, n is the porosity of the clay mass, and S and G_s denote the mass-specific surface area and the specific weight of the solid material, respectively.

Similar to the Kozeny–Carman equation, Towhata et al. (1993a) suggested that the hydraulic conductivity varies with temperature as:

$$\frac{K(T)}{K(T_0)} = \frac{f(T)}{f(T_0)}\frac{g'(T)}{g'(T_0)}\frac{h'(T)}{h'(T_0)} \tag{3.25}$$

where T_0 denotes the reference temperature, f denotes the effects of pore size and shape, g' denotes the effect of void ratio, and h' denotes the property of pore water at the corresponding temperature. However, Towhata et al. (1993a) assumed that "f" and "g'" in Eq. (3.24) were equal, which means the change of hydraulic conductivity can be predicted only by considering the variation of water properties (viscosity and density) with temperature.

Based on the Kozeny–Carman equation, and assuming that the variation of S and G_s can be neglected from the temperature range of 20–90 °C (Towhata et al. 1993a), the hydraulic conductivity of clay at different temperatures T can be represented as

$$K(T) = \frac{C(T)\mu_w(T_0)\rho_w(T_0)(1-n(T_0))^2(n(T))^3}{C(T_0)\mu_w(T)\rho_w(T)(1-n(T))^2(n(T_0))^3}K(T_0) \tag{3.26}$$

Equation (3.26) indicates that the changes in the physical properties of water are not the only ones that influence the changes in the hydraulic conductivity with temperature. As the temperature increases, the thermal effects result in regrouping of the arrangement of clay particles, altering the clay fabric and porosity redistribution (Romero et al. 2001). Pusch and Güven (1990) observed the birth of larger voids between clay particles during heating by using analytical electron microscopy

(AEM). Additionally, the absorbed water may degenerate into free water under thermal loading (Derjaguin et al. 1986). The tortuosity and shape of the flow channels may have changed as a result of the above phenomenon, which is also the case for the values of C and n. Furthermore, the porosity would also be influenced by the thermally induced volume change behavior (Delage et al. 2000).

The predictions of Eqs. (3.25) and (3.26) were compared in Fig. 3.10. The variation of parameters at different temperatures is shown in Table 3.1. The changes in hydraulic conductivity and clay porosity with temperature are measured experimentally by Delage et al. (2000). Keep in mind that the ratio of "f" and "g" in Eq. (3.25) is equal to unity (Towhata et al. 1993a), and the ratio of "C" in Eq. (3.26) is at present assumed equal to unity. Equation (3.26) can better predict the change of hydraulic conductivity with temperature. Equation (3.25) overestimates the values of hydraulic conductivity because it is predicted by considering only the variation of water properties (viscosity and density) with temperature and assuming that clay porosity remains constant.

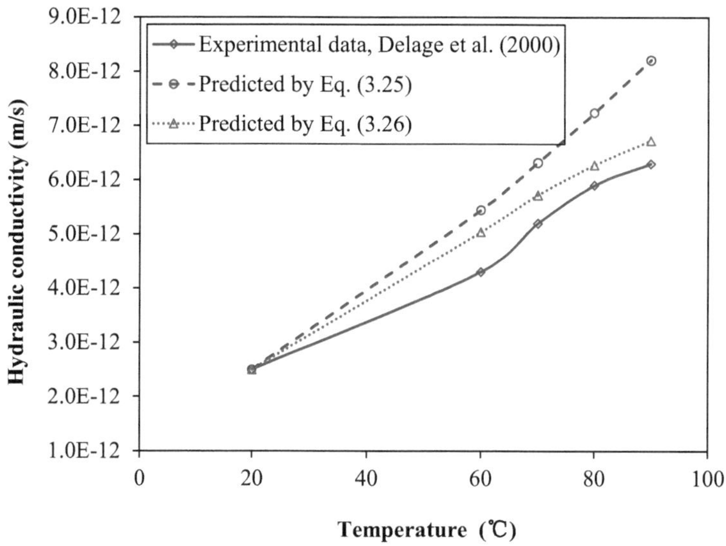

Fig. 3.10 Comparison of the predictions of Eqs. (3.25) and (3.26)

Table 3.1 Variation of parameters at different temperatures in Eqs. (3.25) and (3.26)

Temperature (°C)	20	60	70	80	90
Water viscosity (10^{-3} Pa·s)	1.002	0.6560	0.4688	0.3565	0.3165
Water density (g/cm^3)	0.998	0.983	0.978	0.972	0.965
Clay porosity (%)	39.0	38.3	38.1	37.7	37.2

3.5.2 Hydraulic Conductivity Evolution Model Considering TM-Coupled Effect

As a porous medium, the variation of the void ratio of Boom Clay is controlled by the deformation of the clay skeleton.

$$\Delta e = \Delta\left(\frac{V_p}{V_s}\right) \tag{3.27}$$

where Δe is the void ratio, V_p is the pore volume, and V_s is the solid volume. Assuming that the clay particles are incompressible, the volume strain can be obtained as:

$$\varepsilon_v = \frac{\Delta V_p}{V_s\left(1+e_0\right)} = \frac{V_s \Delta e}{V_s\left(1+e_0\right)} = \frac{\Delta e}{\left(1+e_0\right)} = \frac{e-e_0}{1+e_0} \tag{3.28}$$

where e_0 is the initial void ratio, and e is the void ratio. Based on Eq. (3.28), the porosity can be obtained.

$$n = 1 - \frac{1-n_0}{1+\varepsilon_v} \tag{3.29}$$

where n is the porosity, and n_0 is the initial porosity.

Based on the results of THM-coupled permeability tests in Chap. 2, the changes in viscosity and density of water with temperature are the main factors affecting the change in hydraulic conductivity of Boom Clay with temperature. Take Eq. (3.29) into Eq. (3.24), considering the temperature effect on the hydraulic conductivity and assuming that the K, g, S, and G_s in Eq. (3.24) are invariant, the hydraulic conductivity under various temperatures and stress states can be obtained with the following expression:

$$K\left(T\right) = K\left(T_0\right)\frac{\rho_w\left(T_0\right)\mu_w\left(T_0\right)}{\rho_w\left(T\right)\mu_w\left(T\right)}\left(\frac{\left(\varepsilon_v + n_0\right)^3}{n_0^3\left(1+\varepsilon_v\right)}\right) \tag{3.30}$$

where T_0 is initial temperature, $K(T_0)$ is the hydraulic conductivity at initial temperature, and $\mu_w(T_0)$ and $\rho_w(T_0)$ are the dynamic viscosity and the mass density of water, respectively.

3.6 Numerical Implementation of the Proposed Model in ABAQUS

3.6.1 Coupled Thermo-Hydro-Mechanical Model for Porous Media

The coupled thermo-hydro-mechanical model for deformable saturated porous medium is based on the momentum conservation equation, mass conservation equation, energy conservation equation, constitutive equations, and the Terzaghi's effective stress principle (Terzaghi 1943).

(1) Momentum conservation equation

According to Terzaghi's effective stress principle, the total stress of clay can be decomposed into several parts.

$$\sigma = \sigma' + \alpha_{w} p_{w} \delta = \mathbf{C}^{e}\left(\varepsilon - \varepsilon_{p} - \varepsilon_{c} - \beta \Delta T\right) + \alpha_{w} p_{w} \delta \tag{3.31}$$

where $\boldsymbol{\sigma}$ is the stress tensor, $\boldsymbol{\sigma}'$ is the effective stress tensor, $\mathbf{C}$ is the elastic modulus tensor, $\boldsymbol{\varepsilon}$ is the strain tensor, T is temperature, $\boldsymbol{\beta}$ is the thermal expansion coefficient of the clay skeleton, $\boldsymbol{\varepsilon}_{c}$ is the creep strain tensor, $\boldsymbol{\delta}$ is the Kronecker Delta tensor, p_{w} is the pore water pressure, and $\alpha_{w} = 1$ is the Biot coefficient.

Taking the derivative of both sides of Eq. (3.31) leads to:

$$d\sigma = \mathbf{C}\left(d\varepsilon - d\varepsilon_{p} - d\varepsilon_{c} - \beta dT\right) + d\mathbf{C}^{e}\left(\varepsilon - \varepsilon_{p} - \varepsilon_{c} - \beta \Delta T\right) + \alpha_{w} dp_{w} \delta \tag{3.32}$$

The balance equation of solid skeleton can be formulated as follows:

$$\nabla \cdot \sigma + \mathbf{f} = \mathbf{0} \tag{3.33}$$

where $\mathbf{f} = \rho \mathbf{g}$ is the body force, and ρ is the wet density.

According to the small deformation assumption, the total strain tensor is defined as

$$\varepsilon = \frac{1}{2}\left(\mathbf{u}\nabla + \nabla \mathbf{u}\right) \tag{3.34}$$

where $\mathbf{u}$ is the displacement vector.

Substituting Eq. (3.34) into Eq. (3.33), and adopting the small deformation assumption, the following momentum conservation equation can be obtained:

$$\nabla\cdot\left\{\begin{aligned}&\mathbf{C}:\left[\frac{1}{2}\left(\frac{\mathrm{d}\mathbf{u}}{\mathrm{d}t}\nabla+\nabla\frac{\mathrm{d}\mathbf{u}}{\mathrm{d}t}\right)-\frac{\mathrm{d}\varepsilon_{\mathrm{p}}}{\mathrm{d}t}-\frac{\mathrm{d}\varepsilon_{\mathrm{c}}}{\mathrm{d}t}-\beta\frac{\mathrm{d}T}{\mathrm{d}t}\right]\\&+\left(\frac{\mathrm{d}\mathbf{C}}{\mathrm{d}t}\right):\left[\frac{(\mathbf{u}\nabla+\nabla\mathbf{u})}{2}-\varepsilon_{\mathrm{p}}-\varepsilon_{\mathrm{c}}-\beta\Delta T\right]\\&+\alpha_{\mathrm{w}}\frac{\mathrm{d}p_{\mathrm{w}}}{\mathrm{d}t}\delta+\mathrm{d}\rho\mathbf{g}\end{aligned}\right\}=\mathbf{0} \tag{3.35}$$

(2) Mass conservation equations

The material time derivative can be defined as

$$\frac{\mathrm{d}(*)}{\mathrm{d}t}=\frac{\partial(*)}{\partial t}+\nabla(*)\cdot\mathbf{v} \tag{3.36}$$

in which $\mathbf{v}=\frac{\partial\mathbf{u}}{\partial t}$ is the velocity vector of the solid phase.

The overall quality of the solid phase is

$$M_{\mathrm{s}}=\int_V(1-n)\rho_{\mathrm{s}}\mathrm{d}V \tag{3.37}$$

where ρ_s is the skeleton density.

Taking the material time derivative of M_{s}

$$\frac{\mathrm{d}M_{\mathrm{s}}}{\mathrm{d}t}=\frac{\mathrm{d}}{\mathrm{d}t}\left[\int_V(1-n)\rho_{\mathrm{s}}\mathrm{d}V\right]\approx\int_V\left\{(1-n)\frac{\partial\rho_{\mathrm{s}}}{\partial t}-\rho_{\mathrm{s}}\frac{\partial n}{\partial t}+\nabla\cdot\left[(1-n)\rho_{\mathrm{s}}\mathbf{v}_{\mathrm{s}}\right]\right\}\mathrm{d}V=0 \tag{3.38}$$

Thus, the mass conservation equation of the solid phase is written as

$$(1-n)\frac{\partial\rho_{\mathrm{s}}}{\partial t}-\rho_{\mathrm{s}}\frac{\partial n}{\partial t}+(1-n)\rho_{\mathrm{s}}\nabla\cdot\mathbf{v}_{\mathrm{s}}=0 \tag{3.39}$$

The density of the solid phase is a function of effective stress, water pressure, and temperature:

$$\rho_{\mathrm{s}}=\rho_{\mathrm{s}}\left(p_{\mathrm{w}},\varepsilon_{\mathrm{v}},T\right) \tag{3.40}$$

The time derivative of solid density can be obtained as

$$\frac{\partial\rho_{\mathrm{s}}}{\partial t}=\frac{\rho_{\mathrm{s}}}{K_{\mathrm{s}}}\frac{\partial p_{\mathrm{w}}}{\partial t}+\rho_{\mathrm{s}}\beta\frac{\partial T}{\partial t} \tag{3.41}$$

Inserting Eq. (3.41) into Eq. (3.39), the mass conservation equation for solid phase takes the form:

$$(1-n)\left(\frac{1}{K_{\mathrm{s}}}\frac{\partial p_{\mathrm{w}}}{\partial t}+\rho_{\mathrm{s}}\beta\frac{\partial T}{\partial t}\right)-\frac{\partial n}{\partial t}+(1-n)\frac{\partial \varepsilon_{\mathrm{v}}}{\partial t}=0 \tag{3.42}$$

If neglecting the thermo-osmosis effect, the relative apparent flow velocity of water is governed by Darcy's law

$$\mathbf{v}_{\mathrm{rw}}=\frac{\mathbf{k}}{\mu_{\mathrm{w}}}(\nabla p_{\mathrm{w}}-\rho_{\mathrm{w}}\mathbf{g}) \tag{3.43}$$

where μ_{w} is the dynamic viscosity of water, and $\mathbf{k}$ is the intrinsic velocity vector of water with respect to the deforming solid.

For liquid phase, the integral form of continuous equation

$$M_{\mathrm{w}}=\int_V n\rho_{\mathrm{w}}\mathrm{d}V \tag{3.44}$$

The quality of the water conservation equation (the equation of continuity) can be written as

$$\frac{dM_w}{dt}=\frac{d}{dt}\left(\int_V \rho_w n dV\right)=\int_V\left(\frac{\partial(\rho_w n)}{\partial t}+\nabla\cdot(\rho_w n v_w)\right)dV=0 \tag{3.45}$$

where $\mathbf{v}_{\mathrm{w}} = \mathbf{v}_{\mathrm{wr}} + \mathbf{v}_{\mathrm{s}}$.

$$\frac{\partial \rho_{\mathrm{w}}}{\partial t}=\frac{\rho_{\mathrm{w}}}{K_{\mathrm{w}}}\frac{\partial p_{\mathrm{w}}}{\partial t}+\rho_{\mathrm{w}}\beta_{\mathrm{w}}\frac{\partial T}{\partial t} \tag{3.46}$$

where β_{w} is the thermal expansion coefficient of water.

Taking the material time derivative of M_{w} and substituting Eq. (3.46) into Eq. (3.45) leads to

$$\frac{\mathrm{d}M_{\mathrm{w}}}{\mathrm{d}t}=\int_V\left\{\begin{array}{l} n\left(\frac{\rho_{\mathrm{w}}}{K_{\mathrm{w}}}\frac{\partial p_{\mathrm{w}}}{\partial t}+\rho_{\mathrm{w}}\beta_{\mathrm{w}}\frac{\partial T}{\partial t}\right)+\rho_{\mathrm{w}}\frac{\partial n}{\partial t}\\ +\nabla\cdot\left[\rho_w n\left(\frac{\mathbf{k}}{\mu_{\mathrm{w}}}\nabla p_{\mathrm{w}}-\frac{\mathbf{k}}{\mu_{\mathrm{w}}}\rho_{\mathrm{w}}\mathbf{g}+\mathbf{v}_{\mathrm{s}}\right)\right]\end{array}\right\}\mathrm{d}V \tag{3.47}$$
$$=0$$

From the above equation, we have

$$\frac{n}{K_{\mathrm{w}}}\frac{\partial p_{\mathrm{w}}}{\partial t}+n\beta_{\mathrm{w}}\frac{\partial T}{\partial t}+\frac{\partial n}{\partial t}+\nabla\cdot\left[n\frac{\mathbf{k}}{\mu_{\mathrm{w}}}(\nabla p_{\mathrm{w}}-\rho_{\mathrm{w}}\mathbf{g})\right]+n\frac{\partial \varepsilon_{\mathrm{v}}}{\partial t}=0 \tag{3.48}$$

Summing up the mass conservation equations of the solid phase and pore water yields

$$\left(\frac{n}{K_{\mathrm{w}}}+\frac{1-n}{K_{\mathrm{s}}}\right)\frac{\partial p_{\mathrm{w}}}{\partial t}+\left[n\beta_{\mathrm{w}}+\rho_{\mathrm{s}}\beta(1-n)\right]\frac{\partial T}{\partial t}$$
$$+\nabla\cdot\left[n\frac{\mathbf{k}}{\mu_{\mathrm{w}}}(\nabla p_{\mathrm{w}}-\rho_{\mathrm{w}}\mathbf{g})\right]+\frac{\partial\varepsilon_{\mathrm{v}}}{\partial t}=0 \tag{3.49}$$

(3) Energy conservation equations

Heat conduction follows Fourier's law, and the conduction of heat flux can be represented as

$$\mathbf{q}=\mathbf{q}_{\mathrm{s}}+\mathbf{q}_{\mathrm{w}}=-\left[(1-n)\boldsymbol{\lambda}_{\mathrm{s}}+nS_{\mathrm{r}}\boldsymbol{\lambda}_{\mathrm{w}}\right]\nabla T=-\boldsymbol{\lambda}\nabla T \tag{3.50}$$

in which $\mathbf{q}_{\mathrm{s}}$, $\mathbf{q}_{\mathrm{w}}$ are the heat conduction flux vectors over the solid phase and pore water, respectively; $\boldsymbol{\lambda}_{\mathrm{s}}$, $\boldsymbol{\lambda}_{\mathrm{w}}$ are the thermal conductivity tensors of the solid phase and pore water, respectively; and $\boldsymbol{\lambda}=(1-n)\boldsymbol{\lambda}_{\mathrm{s}}+nS_{\mathrm{r}}\boldsymbol{\lambda}_{\mathrm{w}}$ is the thermal conductivity tensor of the mixture.

The energy conservation equation of solid and pore water can be expressed, respectively, as follows:

$$\rho_{\mathrm{w}}c_{\mathrm{w}}\frac{\partial T}{\partial t}+\mathbf{v}_{\mathrm{rw}}c_{\mathrm{w}}\frac{\partial T}{\partial t}=\rho_{\mathrm{w}}Q_{\mathrm{w}}-\nabla\cdot\mathbf{q}_{\mathrm{w}}+p_{\mathrm{w}}\boldsymbol{\delta}:\nabla\mathbf{v}_{\mathrm{w}} \tag{3.51}$$

$$\rho_{\mathrm{s}}c_{\mathrm{s}}\frac{\partial T}{\partial t}=\rho_{\mathrm{s}}Q_{\mathrm{s}}-\nabla\cdot\mathbf{q}_{\mathrm{s}}+\boldsymbol{\sigma}_{\mathrm{s}}:\nabla\mathbf{v}_{\mathrm{s}} \tag{3.52}$$

in which Q_{w}, Q_{s} are the heat supply of the solid phase and pore water, respectively.

The energy conservation equation for the mixture can be summarized as

$$\rho c_{\mathrm{v}}\frac{\partial T}{\partial t}+\mathbf{v}_{\mathrm{rw}}c_{\mathrm{w}}\frac{\partial T}{\partial t}=\rho Q-\nabla\cdot\mathbf{q}+\boldsymbol{\sigma}:\nabla\mathbf{v} \tag{3.53}$$

in which $\rho Q=\rho_{\mathrm{w}}Q_{\mathrm{w}}+\rho_{\mathrm{s}}Q_{\mathrm{s}}$, $\rho c_{\mathrm{v}}=\rho_{\mathrm{w}}c_{\mathrm{w}}+\rho_{\mathrm{s}}c_{\mathrm{s}}$.

Introducing Darcy's law and Fourier's law into Eq. (3.53) and neglecting mechanical energy conversion of pore water, we arrive at

$$\rho c_{\mathrm{v}}\frac{\partial T}{\partial t}+\rho_{\mathrm{w}}c_{\mathrm{w}}\nabla\cdot\left[\frac{\rho_{\mathrm{w}}\mathbf{k}}{\mu_{\mathrm{w}}}(\nabla p_{\mathrm{w}}+\rho_{\mathrm{w}}\mathbf{g})\right]\nabla T$$
$$-\nabla\cdot(\boldsymbol{\lambda}\nabla T)=\rho Q+\nabla\cdot\left(\boldsymbol{\sigma}\frac{\mathrm{d}\mathbf{u}}{\mathrm{d}t}\right)+\rho\mathbf{g}\cdot\frac{\mathrm{d}\mathbf{u}}{\mathrm{d}t} \tag{3.54}$$

where c_{v} is the heat capacity of rock skeleton.

(4) THM-coupled FEM formulation in ABAQUS.

The momentum conservation equations, the mass conservation equation, and the energy conservation equation compose the THM coupling governing equations. The basic unknown variables for these equations are displacement **u**, pore water pressure p_w, temperature T, and porosity n. The basic problem is to find a solution in the space domain V of the medium during a time period $[0, t]$.

The initial conditions ($t = 0$) are as follows:

$$\begin{cases} \mathbf{u}(0) = \mathbf{u}_0 \\ p_w(0) = p_{w0} \\ T(0) = T_0 \\ n(0) = n_0 \end{cases} \tag{3.55}$$

The following boundary conditions should be satisfied:

On the stress boundaries, $\boldsymbol{\sigma} \cdot \mathbf{n} = \hat{\boldsymbol{\sigma}}$

On the displacement boundaries, $\mathbf{u} = \hat{\mathbf{u}}$

On the pore pressure boundaries, $p_w = \hat{p}_w$

On the water flux boundaries, $q_w = -\dfrac{\mathbf{k}}{\mu_w}(\nabla p_w - \rho_w \mathbf{g}) \cdot \mathbf{n} = \hat{q}_w$

On the temperature boundaries, $T = \hat{T}$

On the heat flux boundaries, $q_T = -\lambda \nabla T \cdot \mathbf{n} = \hat{q}_T$

On the heat convection boundaries, $-\lambda \nabla T = h(T_\infty - T)$

Take the shape functions to

$$\begin{cases} \mathbf{u} = \mathbf{N}\mathbf{u}^e \\ p_w = \bar{\mathbf{N}}\mathbf{p}_w^e \\ \Delta T = \bar{\mathbf{N}}\Delta\mathbf{T}^e \end{cases} \tag{3.56}$$

where $\mathbf{u}^e$, $\mathbf{p}_w{}^e$, and $\Delta\mathbf{T}^e$ are the nodal displacement matrix, the nodal pore water vector, and the nodal temperature increment vector, respectively, $\mathbf{N} = [N_1\mathbf{I}\ \ N_2\mathbf{I}\ \ \ldots\ \ N_m\mathbf{I}]$ is the shape function matrix, $N_1, N_2 \ldots N_m$ are the shape functions, and $\mathbf{I}$ is the unit matrix, $\bar{\mathbf{N}} = [\mathrm{N}_1 \quad \mathrm{N}_2 \quad \ldots \quad \mathrm{N}_m]$.

By applying the Galerkin method to operate the spatial discretization and introducing the boundary conditions into the governing equations, the numerical matrix equations can be obtained.

$$\mathbf{A}_1\dot{\mathbf{X}}^e + \mathbf{A}_2\mathbf{X}^e + \mathbf{Y} = 0 \tag{3.57}$$

in which

$$\mathbf{X}^e = \begin{Bmatrix} \mathbf{u}^e \\ \mathbf{p}_w^e \\ \mathbf{T}^e \end{Bmatrix} \tag{3.58}$$

The time discretization of the governing equations can be expressed as

$$\dot{\mathbf{X}}^{e} = \frac{\mathbf{X}_{n+1}^{e} - \mathbf{X}_{n}^{e}}{\Delta t} \tag{3.59}$$

$$\mathbf{X}^{e} = (1-\eta)\mathbf{X}_{n+1}^{e} - \eta\mathbf{X}_{n}^{e} \tag{3.60}$$

$$\mathbf{Y}^{e} = (1-\eta)\mathbf{Y}_{n+1}^{e} - \eta\mathbf{Y}_{n}^{e} \tag{3.61}$$

Substitution of Eqs. (3.59)–(3.61) into Eq. (3.58) results in

$$\frac{\mathbf{A}_1\left(\mathbf{X}_{n+1}^{e} - \mathbf{X}_{n}^{e}\right)}{\Delta t} + \mathbf{A}_2\left[(1-\eta)\mathbf{X}_{n+1}^{e} - \eta\mathbf{X}_{n}^{e}\right] + \left[(1-\eta)\mathbf{Y}_{n+1}^{e} - \eta\mathbf{Y}_{n}^{e}\right] = 0 \tag{3.62}$$

A staggered solution scheme with sequentially coupled solutions was taken by many researchers. The uncoupled heat transfer analysis is performed as

$$\mathbf{K}_{\mathrm{TT}}\Delta\mathbf{T} = \Delta\mathbf{R}_{\mathrm{T}} \tag{3.63}$$

Then the hydro-mechanical analysis is conducted.

$$\begin{bmatrix} \mathbf{K}_{\mathrm{uu}} & \mathbf{K}_{\mathrm{up}} \\ \mathbf{K}_{\mathrm{pu}} & \mathbf{K}_{\mathrm{pp}} \end{bmatrix} \begin{Bmatrix} \Delta\mathbf{U} \\ \Delta\mathbf{P} \end{Bmatrix} = \begin{Bmatrix} \Delta\mathbf{R}_{\mathrm{u}} \\ \Delta\mathbf{R}_{\mathrm{p}} \end{Bmatrix} - \mathbf{K}_{\mathrm{uT}}\Delta\mathbf{T} - \mathbf{K}_{\mathrm{pT}}\Delta\mathbf{T} \tag{3.64}$$

The porosity is updated finally.

Establishing Eq. (3.62) for all elements and assembling element governing matrices, the following global governing matrix representation of the coupled equations can be obtained.

$$\begin{bmatrix} \mathbf{K}_{\mathrm{uu}} & \mathbf{K}_{\mathrm{up}} & \mathbf{K}_{\mathrm{uT}} \\ \mathbf{K}_{\mathrm{pu}} & \mathbf{K}_{\mathrm{pp}} & \mathbf{K}_{\mathrm{pT}} \\ \mathbf{K}_{\mathrm{Tu}} & \mathbf{K}_{\mathrm{Tp}} & \mathbf{K}_{\mathrm{TT}} \end{bmatrix} \begin{Bmatrix} \Delta\mathbf{U} \\ \Delta\mathbf{P} \\ \Delta\mathbf{T} \end{Bmatrix} = \begin{Bmatrix} \Delta\mathbf{R}_{\mathrm{u}} \\ \Delta\mathbf{R}_{\mathrm{p}} \\ \Delta\mathbf{R}_{\mathrm{T}} \end{Bmatrix} \tag{3.65}$$

ABAQUS solves these equations in a fully coupled manner.

3.6.2 Secondary Development of the Proposed Model in ABAQUS

ABAQUS provides a series of subroutines programed in FORTRAN for the users. The USDFLD subroutine is designed to update the model parameters at all material points of elements at the start of the increment in an explicit way. It can be used to introduce the evolution law of material properties, such as friction angle, cohesion, and Young's modulus. These parameters are defined as functions of field variables.

The calling processes of USDFLD subroutine in ABAQUS code are:

Step 1: The main code of ABAQUS provides stress, total strain, total strain increment, temperature, and time increment at the start of the increment.
Step 2: Update damage variables, Young's modulus, friction angle, and cohesion based on the hardening law and damage evolution law. Modify the model parameters by using USDFLD subroutine.
Step 3: Calculate the elastic trial stress. Perform loading/unloading check and update the elastic and plastic strain, stress, and the consistent tangent stiffness with the constitutive integration algorithm.
Step 4. Return to the main code of ABAQUS. Perform equilibrium iteration. If the calculation converges, start the next increment. Otherwise, go to Step 1 with a smaller time increment until the calculation converges.

The flow chart of the subroutine is shown in Fig. 3.11.

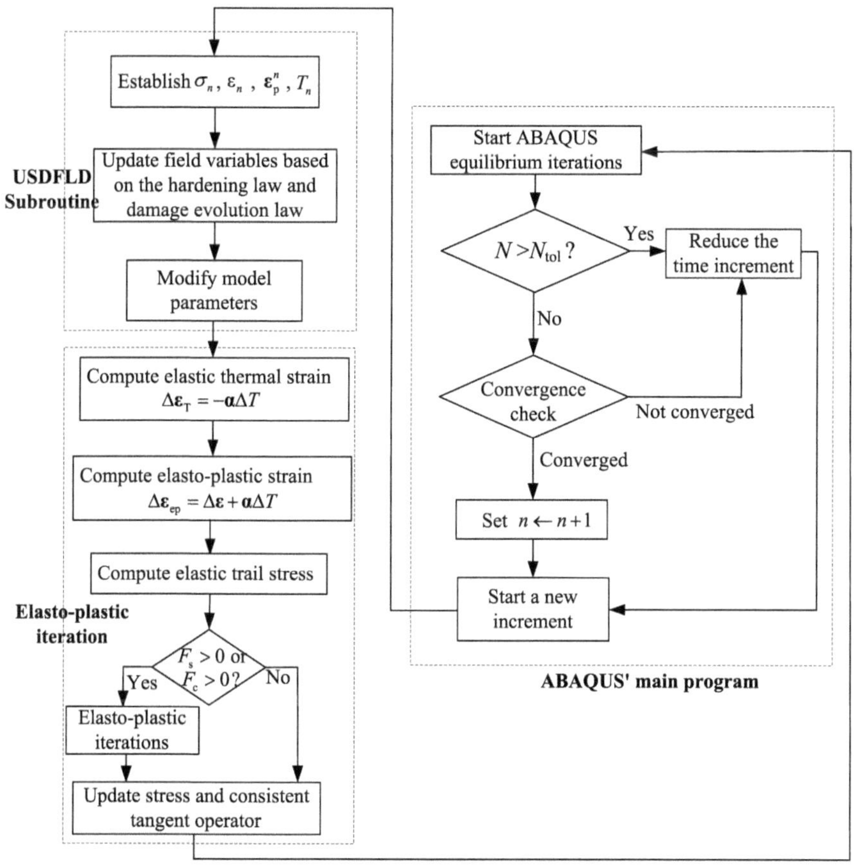

Fig. 3.11 Flow chart for the secondary development of thermo-hydro-mechanical-damage model

3.7 Parameters Inversion from Test Results

3.7.1 Basic THM Parameters and FEM Model

After a careful examination of the experimental and back analysis data from in situ wave velocities of seismic tests, in situ heating tests, and laboratory tests Chen et al. (2011), suggested adopting the anisotropic parameters. The basic THM parameters of Boom Clay are presented in Table 3.2. When considering transverse isotropy, higher values of thermal conductivity, hydraulic conductivity, Poisson ratio, and Young's modulus in the horizontal plane are used, compared to those in the vertical plane.

3.7.2 Damage and Plastic Parameters Determination by Back Analysis

Based on the triaxial test results of Boom Clay in Chap. 2, back analysis has been performed to determine the parameters of the proposed model. For the back analysis problem, to improve its reliability and to get the minimum errors between the measured and predicted values within a certain error range, the measured data and inversion models should be combined. Accordingly, the problem of back analysis can be considered an optimization problem. The Nelder–Mead method, which is the most popular direct search method (Jia 2009; Yu 2010), is thus employed here.

Table 3.2 Basic THM parameters for Boom Clay

Parameter	Value
Young's modulus/MPa	$E_{h0} = 1400$, $E_{v0} = 700$
Cohesion/MPa	$c_0 = 0.4$
Friction angle/°	$\varphi_0 = 15$
Hydraulic conductivity/m s^{-1}	$K_h(T_0) = 6 \times 10^{-12}$, $K_v(T_0) = 3 \times 10^{-12}$
Poisson's ratio	$\nu_{hh} = 0.25$, $\nu_{vh} = 0.125$
Dry density/kg m^{-3}	1640
Porosity/%	$n_0 = 39$
Preconsolidation pressure/MPa	$p_{c0} = 5.0$
Specific heat capacity/J kg^{-1} K^{-1}	740
Thermal conductivity/W m^{-1} K^{-1}	$\lambda_h = 2.33$, $\lambda_v = 1.77$
Thermal expansion coefficient/K^{-1}	1×10^{-5}
Density of pore water/kg m^{-3}	$\rho_w (T_0) = 999$
Viscosity of pore water/Pa s	$\mu_w (T_0) = 1.101 \times 10^{-3}$
Specific heat capacity of pore water/J kg^{-1} K^{-1}	4200
Thermal conductivity of pore water/W m^{-1} K^{-1}	0.63
Thermal expansion coefficient of pore water/K^{-1}	4.5×10^{-4}

Figure 3.12 shows the 3D finite element model for Boom Clay sample with the dimension of Φ 38 × 76 mm. The bottom of the model is fixed in the axial direction, and then the bottom center of the model is fixed in both the axial and radial directions. The boundary conditions are the same as in the test. The undrained triaxial compression test is simulated numerically by the THMD coupled model presented.

Stress and pore water pressure are the main test data that can be used in the back analysis. To use both data, a conjoint back analysis model is established. A reasonable relative weight ratio is given in the model to reflect the weight of the different physical quantities. For a given axial strain, the stress and pore water error between numerical calculations and tests are used in the least squares method. Back analysis needs to minimize the objective function. Therefore, the objective function of the conjoint back analysis is

$$(1-\lambda)\varphi_1 + \lambda\varphi_2 \rightarrow \min \tag{3.66}$$

The objective function based on stress can be expressed as

$$\varphi_1\left(\beta_{\mathrm{M}},\beta_{\mathrm{T}},\bar{D}_p,\beta_{\mathrm{p}},h_{\mathrm{p0}},\cdots\right) = \sum_{k=1}^{l}\left(\sigma_{ij}^{\mathrm{calk}} - \sigma_{ij}^{\mathrm{testk}}\right)^2 \rightarrow \min \tag{3.67}$$

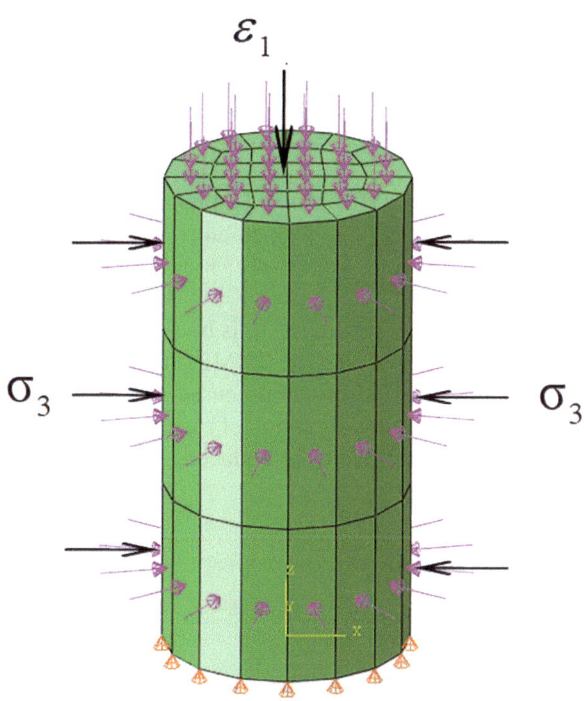

Fig. 3.12 Schematic diagram of the finite element model for test simulations

Table 3.3 Damage and plastic parameters of the THMD model

Parameter	Value	Parameter	Value
β_M	0.035	$\overline{e}_{oe}$	0.05
$\beta_T/°C^{-1}$	0.006	β_p	2380
a	2.89	$\bar{D}_p$	0.35
h_{p0}	0.15	b_p	50.89
c_p	8.4		

where $\sigma_{ij}^{\text{test}}$ are the test data corresponding to axial strains, $\sigma_{ij}^{\text{calk}}$ are the back analysis calculated value, and l is the number of test data used for back analysis.

The objective function based on the pore water pressure can be expressed as

$$\varphi_2\left(\beta_{\text{M}},\beta_{\text{T}},\bar{D}_p,\beta_{\text{p}},h_{\text{p0}},\cdots\right)=\sum_{k=1}^{l}\left(p_{wij}^{\text{calk}}-p_{wij}^{\text{testk}}\right)^2\rightarrow\min \tag{3.68}$$

where p_{wij}^{test} are the test data corresponding to axial strains, and p_{wij}^{calk} are the back analysis calculated value, and l is the number of test data used for back analysis.

The calibration of parameters is done using the test data obtained in Chap. 2. Table 3.3 shows the back analysis results for the undetermined parameters.

3.8 Numerical Simulation on THM-Coupled Undrained Triaxial Compression Tests

Chapter 2 describes a series of THM-coupled undrained triaxial compression tests under various confining pressures and temperatures. The data from THM-coupled undrained triaxial compression tests were used to validate the proposed THMD model.

The comparison between the simulation (this book; Gong 2015) and the experimental results is shown in Fig. 3.13. The THMD model proposed here can well reproduce the four stages of the stress–strain curve of Boom Clay, especially the plastic flow stage. Furthermore, the evolution of excess pore water pressure during the triaxial compression test can also be reproduced.

Fig. 3.13 Comparison between the simulation (this book; Gong 2015) and the experimental results. (**a**) Simulation of THM_{47402}; (**b**) simulation of THM_{47602}

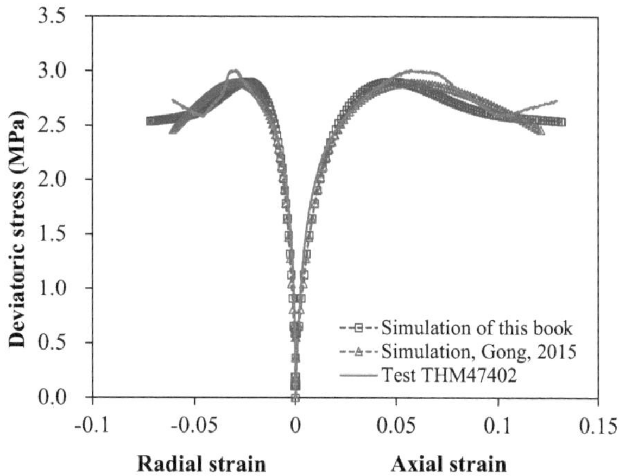

(a) Simulation of THM_{47402}

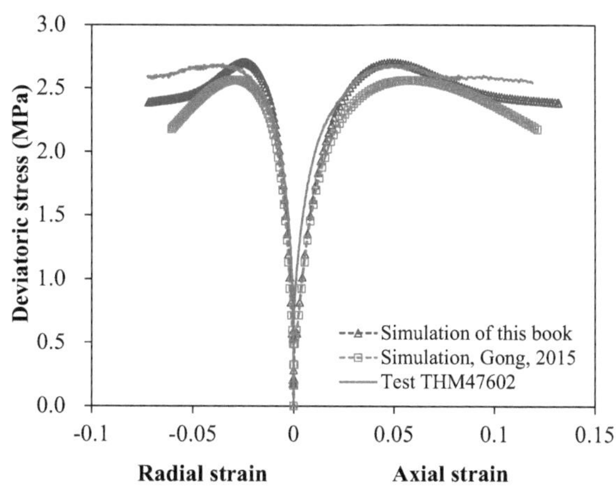

(b) Simulation of THM_{47602}

Figure 3.14 shows the comparison of simulated results and test data of THM-coupled triaxial compression tests. It is indicated that the developed model can describe the hardening, softening, and plastic flow phenomenon in the triaxial compression tests. Moreover, the temperature effect on the strength and excess pore water pressure was reproduced; see Fig. 3.15.

Fig. 3.14 Simulation of stress–strain behavior of THM-coupled undrained triaxial compression tests. (**a**) p'_0 = 1.5 MPa; (**b**) p'_0 = 2.5 MPa; (**c**) p'_0 = 3.5 MPa

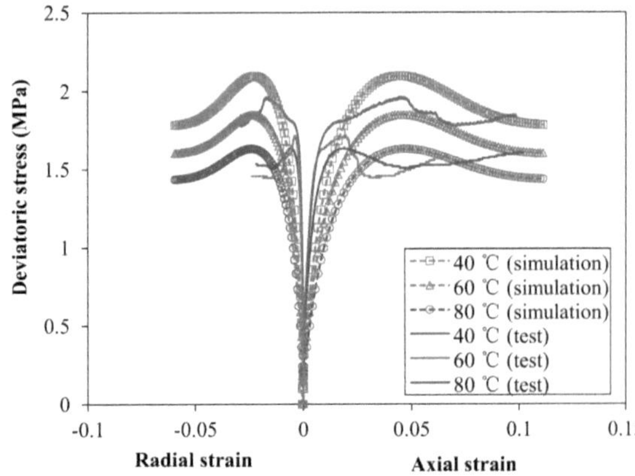

(a) p'_0= 1.5MPa

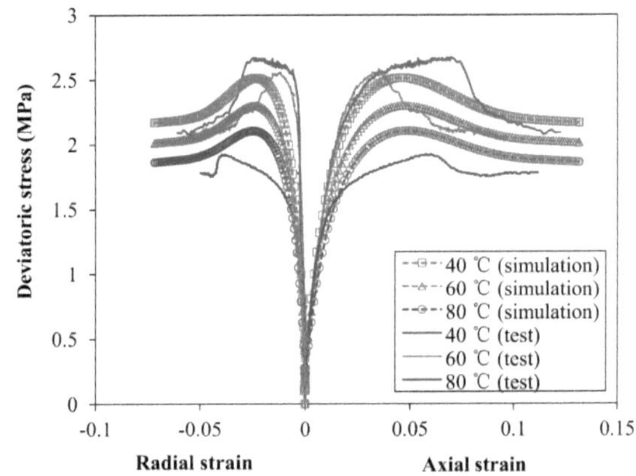

(b) p'_0= 2.5MPa

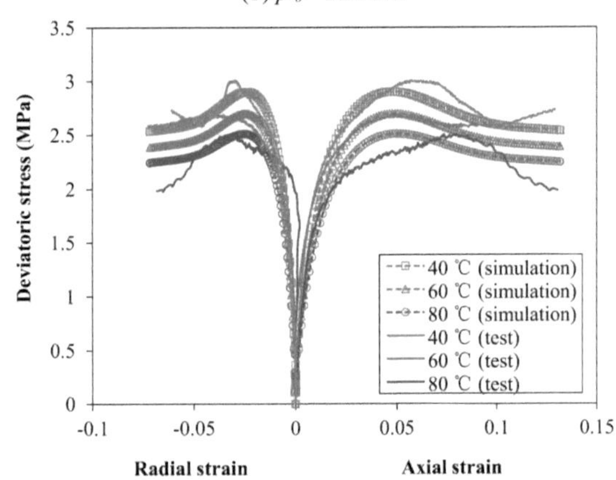

(c) p'_0= 3.5MPa

Fig. 3.15 Simulation of excess pore pressure of THM-coupled undrained triaxial compression tests. (**a**) p'_0 = 1.5 MPa; (**b**) p'_0 = 2.5 MPa; (**c**) p'_0 = 3.5 MPa

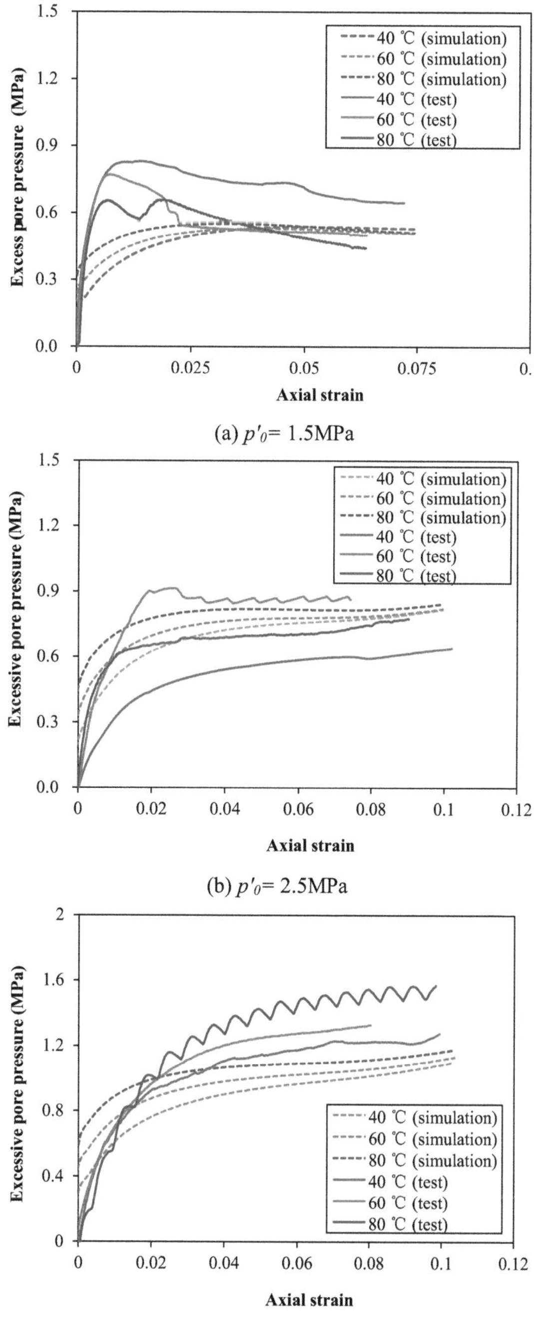

3.9 Conclusions

Based on the Drucker–Prager Cap model, a transversely isotropic thermo-hydro-mechanical-damage (THMD) model was proposed to describe the hydro-mechanical behavior of Boom Clay with temperatures. The elastic damage evolution, the thermal damage evolution, the plastic damage evolution, the plastic hardening law, and the hydraulic conductivity evolution model were introduced based on the results of the THM-coupled triaxial tests and THM-coupled permeability tests of Boom Clay. The proposed model was secondarily developed and implemented in the ABAQUS finite element code through the subroutine USDFLD. Damage and plastic parameters were determined through back analysis using the test results obtained in Chap. 2. The good agreement between simulated results and test data highlights the THM-coupled behavior of Boom Clay and validates the proposed model. The THMD model can well reproduce the four stages of the stress–strain curve of Boom Clay, especially the plastic flow stage. Furthermore, the evolution of excess pore water pressure during the triaxial compression test can also be reproduced.

Chapter 4
Experimental Study on THM-Coupled Creep Behavior of Boom Clay

4.1 Introduction

The mechanical behavior of geomaterials shows not only elasticity and plasticity but also the nature of time-related properties. A time-dependent deformation occurring in a material subject to load for a prolonged period is called creep (Murakami 2012). In a narrower sense, creep means a time-dependent deformation caused by a constant load, which has a significant impact on the stability of underground structures, such as nuclear waste storage facilities, power plants, and tunnels (Kontogianni et al. 2006; Tsai et al. 2008). Although the creep behavior of clay rock has been widely investigated (Gasc-Barbier et al. 2004; Naumann et al. 2007; Fabre and Pellet 2006; Zhang et al. 2007), the experimental studies on the creep behavior of Boom Clay were still limited (De Bruyn et al. 1987; Cui et al. 2009), especially on the creep behavior under the THM coupling effect (Giraud and Rousset 1996). The creep behavior of Boom Clay under the THM coupling effect is very complicated, and many mechanisms involved are still to be understood. Thus, to gain insight into these aspects of behavior, a systematic research program has been carried out on original Boom Clay samples to investigate changes in long-term THM properties.

4.2 HM Creep Tests

4.2.1 Test Methodology and Procedures

The preparation of samples and SBCW, as well as test equipment, is described in Sect. 2.2.1. Table 4.1 gives the basic physical information of the test samples for HM creep behavior study.

W. Chen et al., *Thermo-Hydro-Mechanical Coupling in Clay Rock*,
https://doi.org/10.1007/978-981-95-2562-1_4

Table 4.1 Physical information for the samples tested

Samples	Diameter (mm)	Height (mm)	Density (g/cm^3)	Dry density (g/cm^3)	Water content (%)	Void ratio	Saturation degree	Fluid in contact
TCP1	38.10	74.00	2.04	1.65	24.0	0.66	0.98	SBCW
TCP2	37.83	76.53	2.02	1.63	23.9	0.68	0.96	SBCW
TCP3	37.50	75.77	2.02	1.63	24.0	0.68	0.96	SBCW

Table 4.2 Test procedures during the creep phase after saturation

Samples	Confining pressure σ_3 (MPa)	Water pressure p_l (MPa)	Deviatoric stress $(\sigma_1 - \sigma_3)$ (MPa)	Drained conditions
TCP1	4.7	2.2	0.5, 1.0, 1.5,	Undrained
	4.7	0.0	2.0, 2.5, 3.0, 3.5, 4.0	Drained
TCP2	4.7	2.2	0.5	Undrained
TCP3	4.5	2.0	0.5, 1.0, 1.5, 2.0, 2.5	Drained

Note: In this book, compression is positive, and tension is negative

Before the creep stage, the samples were saturated first. The saturation procedures are similar to the THM undrained triaxial compression tests (see Sect. 2.2.1.4). After saturation, the confining pressure and the back pressure increased to 4.7 MPa and 2.2 MPa, respectively (close to the in situ stress state). After the loading preparations above, the creep phase, including several deviatoric stress steps, began, with each step being 2 months long (see Table 4.2). According to the creep phase, there were three reasons behind the selection of the stress path and test duration. First, for all the samples, we imposed subsequent deviatoric stress steps shown in Table 4.2 to characterize the creep behavior at low stress levels and to obtain information on a possible creep threshold. Second, for the first and the third sample, to compare the influence of the HM coupling effect on creep behavior, the first three deviatoric stress steps and effective stress for both samples are the same except for the drained condition. Third, for the drained creep phase of the first sample, the deviatoric stress was applied stepwise from 2.0 to 4.0 MPa on the samples after the undrained creep phase to characterize creep behavior at relatively high deviatoric stress levels and high confining stress.

4.2.2 *Test Results and Discussion*

4.2.2.1 Creep Deformation

The evolution of the confining pressure, back pressure, strain, and pore water pressure during saturation is similar to the THM undrained compression tests.

Figure 4.1 shows the environmental temperature variation controlled by the air conditioner for the triaxial creep tests (TCP1, TCP2, and TCP3). The longest triaxial creep test (TCP1) was conducted for about 1.5 years.

Figures 4.2 and 4.3 show the variation of deviatoric stress ($\sigma_1 - \sigma_3$), strain, and pore water pressure test for samples TCP1, TCP2, and TCP3 during the creep stage. There are some problems with the radial deformation transducer of test TCP2 and TCP3 during the test, so the radial deformation and volumetric deformation of these samples are absent in the figures. The figures demonstrate that:

- The axial strain and volumetric strain led to an instantaneous increment when the creep load increased, and before the creep deformation began. The volume of the sample endured compression first and then began to swell gradually.
- According to the increasing creep load, the instantaneous deformation caused by the same load increment tended to decrease with increasing creep load. Pore water pressure increased gradually and tended to stabilize gradually.
- The variation of hydrostatic pressure is nearly the same as that of pore pressure in about 2000 hours, and it shows the saturation process of the samples.
- Sample TCP1 (Fig. 4.4a) failed at the loading step of 4.0 MPa under the effective confining pressure of 4.7 MPa after one and a half years. Sample TCP3 (Fig. 4.4b) failed at the loading step of 2.5 MPa under the effective confining pressure of 2.5 MPa after 8 months.

To facilitate our analysis, an offset was applied to make the curves have the same origin of time. Each step starts at $t = 0$ and $\varepsilon_{\text{initial}} = 0$. Figures 4.5 and 4.6 show the influence of applied deviatoric stress on the strain–time curve. It is important to see that all the curves in Fig. 4.5 (Fig. 4.6) are relative to the sample TCP1 (TCP3), and

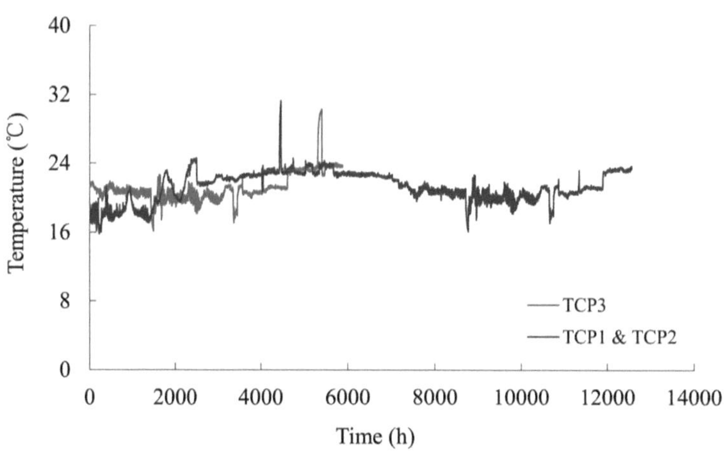

Fig. 4.1 Temperature variation during triaxial creep test

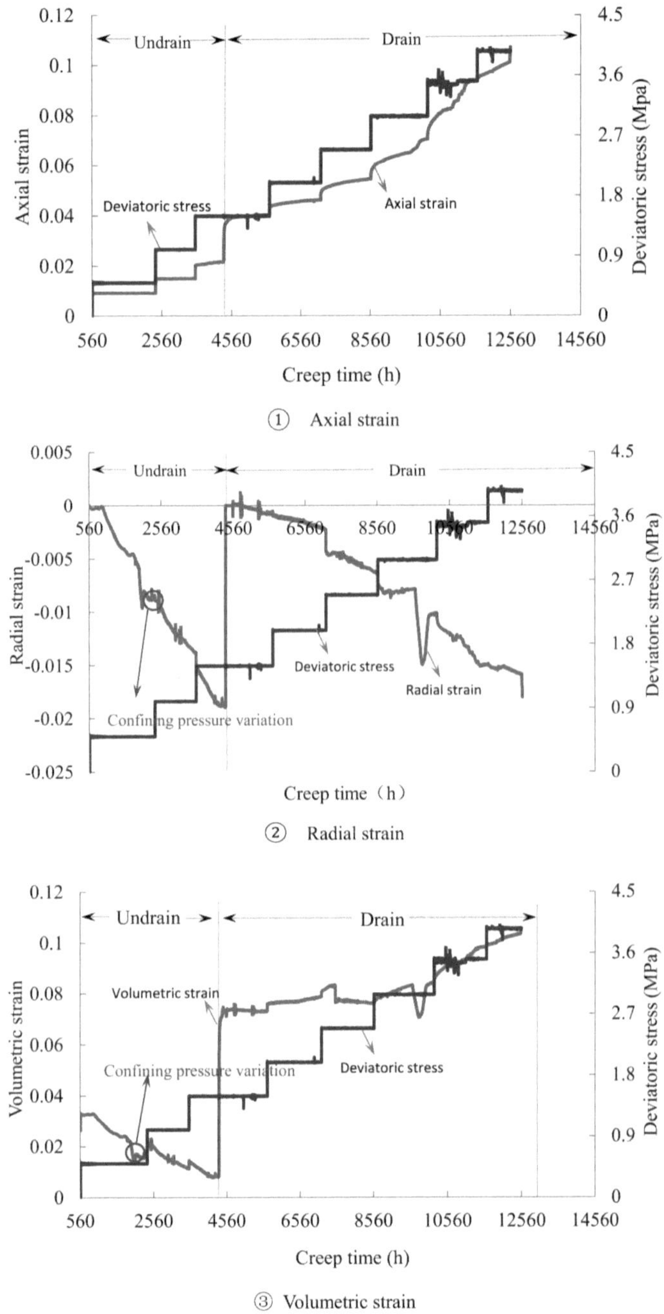

(a) Sample TCP1

Fig. 4.2 Stress–strain variation during creep phase. (**a**) Sample TCP1. (**b**) Sample TCP2. (**c**) Sample TCP3

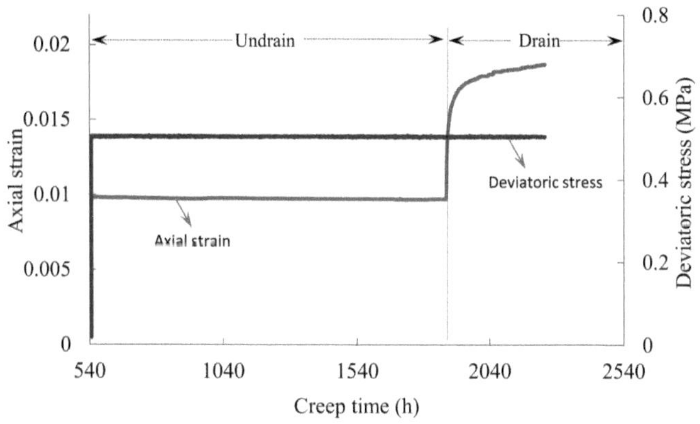

(b) Sample TCP2

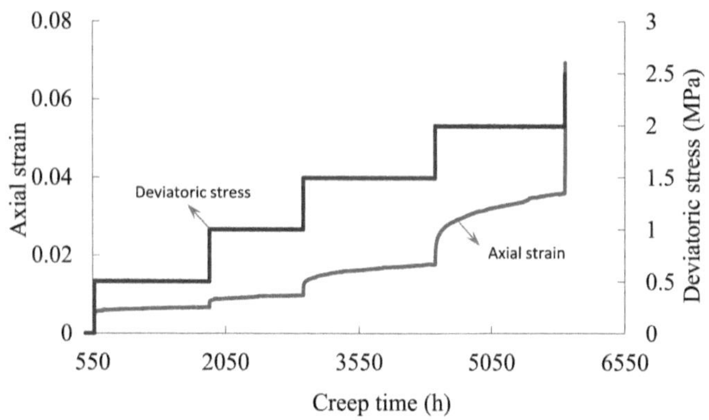

(c) Sample TCP3

Fig. 4.2 (continued)

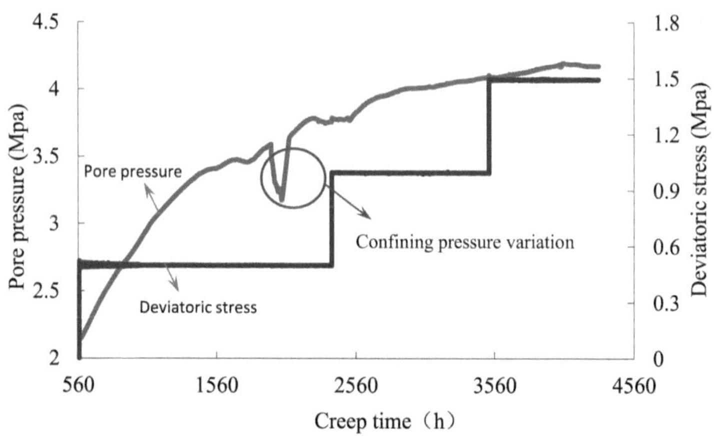

Fig. 4.3 Pore pressure variation during creep phase for sample TCP1

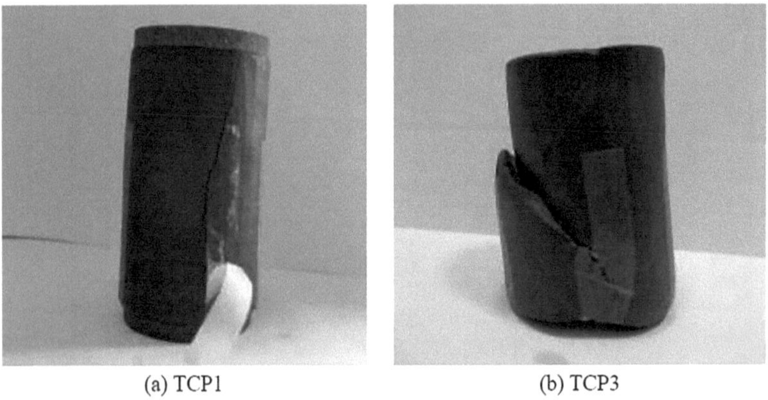

Fig. 4.4 Failed samples for triaxial creep tests. (**a**) TCP1; (**b**) TCP3

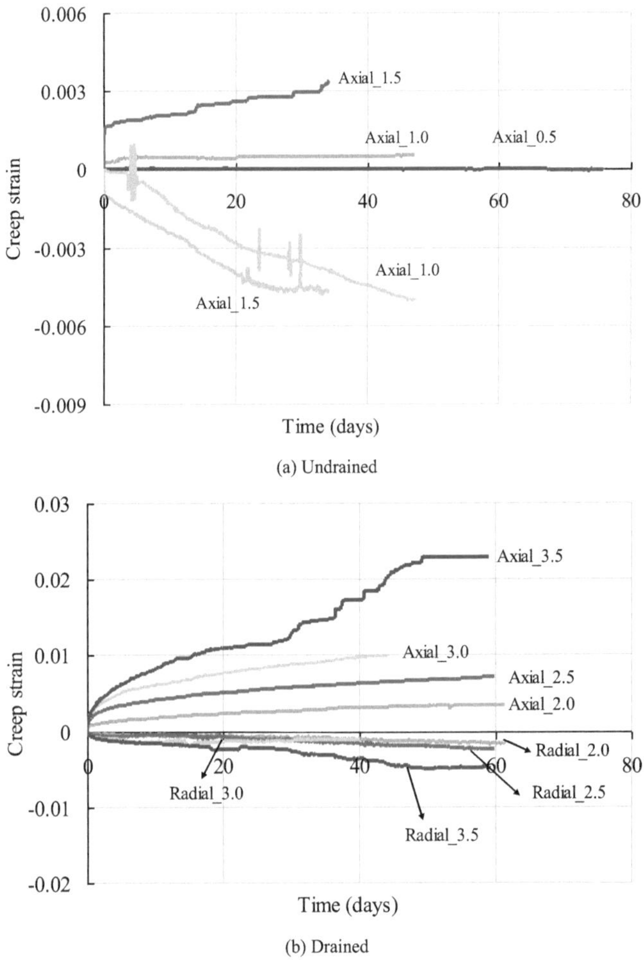

Fig. 4.5 Comparison of creep strains for sample TCP1. (**a**) Undrained; (**b**) drained

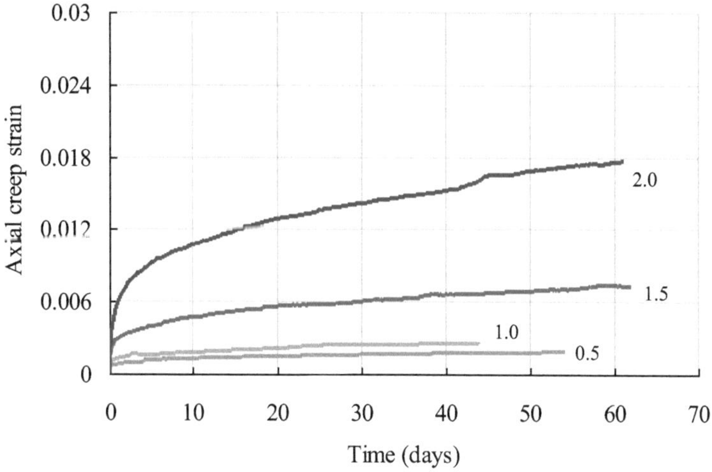

Fig. 4.6 Comparison of creep strains for sample TCP3

each curve corresponds to a different deviatoric stress step. Figures 4.5 and 4.6 clearly show the following:

- Almost no axial creep exists for sample TCP1 and TCP2 under 1.0 MPa deviatoric stress ($\sigma_3 = 4.7$ MPa, $p_l = 2.2$ MPa, undrained). However, a very small creep exists under a 1.0 MPa deviatoric stress for sample TCP3 ($\sigma_3 = 4.5$ MPa, $p_l = 2.0$ MPa , drained). Therefore, if a threshold exists, it is in the vicinity of 1.0 MPa. This value is the same as the threshold value obtained by Giraud (Giraud and Rousset 1996) ($\sigma_3 = 5.0$ MPa, $p_l = 2.0$ MPa, undrained). Supplemental long-term creep experiments are needed to support this finding.
- Compare the axial creep strain at low stress levels from 0.5 to 1.5 MPa between the two samples under different drained conditions. It shows that the creep deformation is more significant under drained conditions.
- Measurements indicate that the sample significantly swelled in the radial direction under undrained conditions. The increase in effective confining pressure weakened the swelling when the test continued under drained conditions.

4.2.2.2 Creep Rate

To better understand the long-term behavior of Boom Clay, the creep strain rates for sample TCP1 were calculated and are shown in Figs. 4.7 and 4.8. The creep rate in the drained condition was found to be smaller than that of the undrained condition because of the larger effective confining pressure. It was also found that the strain rate depends on deviatoric stresses. Similar results were reported on Bure clay, e.g., by Gasc-Barbier et al. (2004) and Opalinus clay, e.g., Naumann et al. (2007).

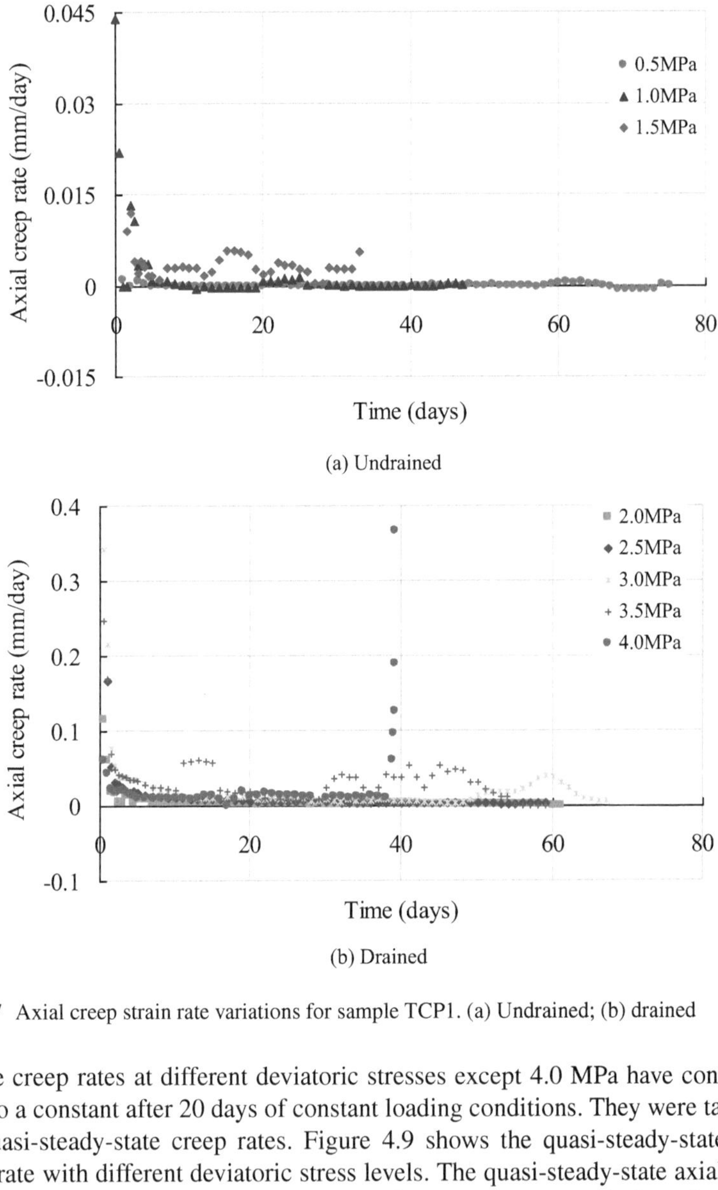

Fig. 4.7 Axial creep strain rate variations for sample TCP1. (a) Undrained; (b) drained

The creep rates at different deviatoric stresses except 4.0 MPa have converged fully to a constant after 20 days of constant loading conditions. They were taken as the quasi-steady-state creep rates. Figure 4.9 shows the quasi-steady-state axial creep rate with different deviatoric stress levels. The quasi-steady-state axial creep rates increase exponentially as the deviatoric stress increases. At low deviatoric stresses (0.5–1.5 MPa), the quasi-steady-state axial creep rate increases from about 0 to 0.003 mm/day. At relatively higher deviatoric stresses of 2.0 and 2.5 MPa, the total stress field of sample TCP1 is the same as sample TCP3, but the effective confining pressures are different; therefore, the quasi-steady-state axial creep rates are different under different HM-coupled conditions. Specifically, the rate was about 0.0085 mm/day under the deviatoric stress of 2.0 MPa for sample TCP3, which is much larger than that of sample TCP1 with 0.002 mm/day.

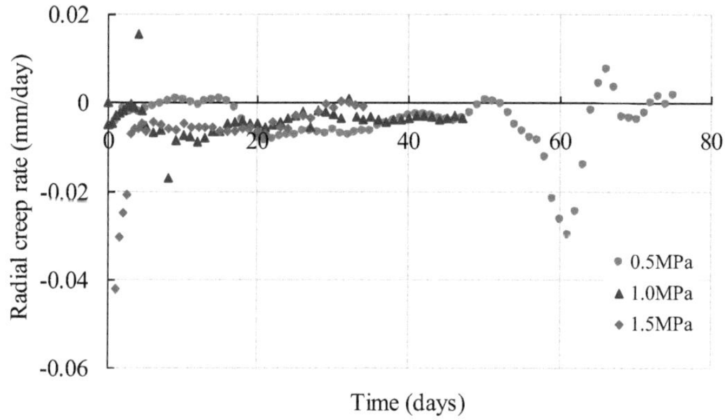

(a) Undrained

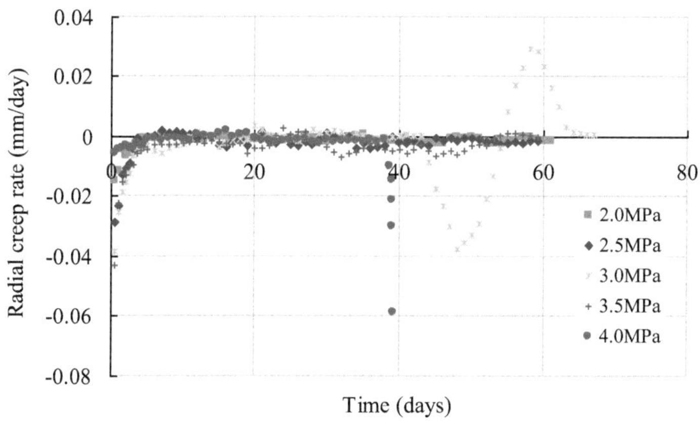

(b) Drained

Fig. 4.8 Radial creep rate variations for sample TCP1. (**a**) Undrained; (**b**) drained

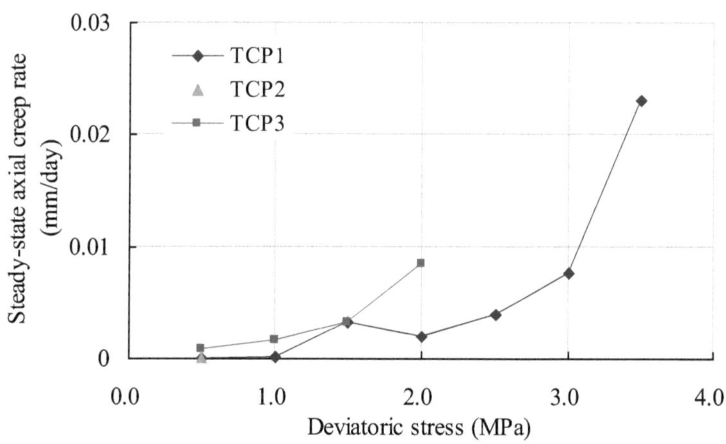

Fig. 4.9 Quasi-steady-state axial creep rate as a function of applied stress for sample TCP1

Numerical simulation (Yu 2010) shows that the Mises stress is around 0.8–1.25 MPa at the location of the in situ measurements (ring 15–105) for the connecting gallery at Hades. Therefore, from the view of long-term engineering stability, the creep under low deviatoric stress is much more concerning. The in situ diameter reduction rates are shown in Fig. 4.10. These in situ diameter reduction rates are the combination of HM-coupled creep behavior of the host rock. It can be observed that fast in situ diameter reduction rates occurred at the beginning, and then the rates soon decreased. This is similar to the steady creep state after 5 years of field monitoring.

Two kinds of in situ quasi-steady-state diameter reduction rates, one calculated from the mean average of the period from 1992 to 2001, and the other calculated from 2001 to 2006, are shown in Fig. 4.11. The former calculated rates are generally larger than the latter, and this shows that the creep behavior weakened during the

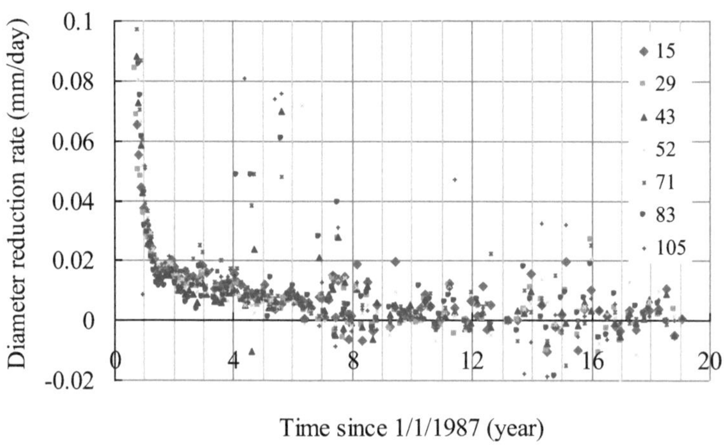

Fig. 4.10 In situ diameter reduction rate for different lining rings

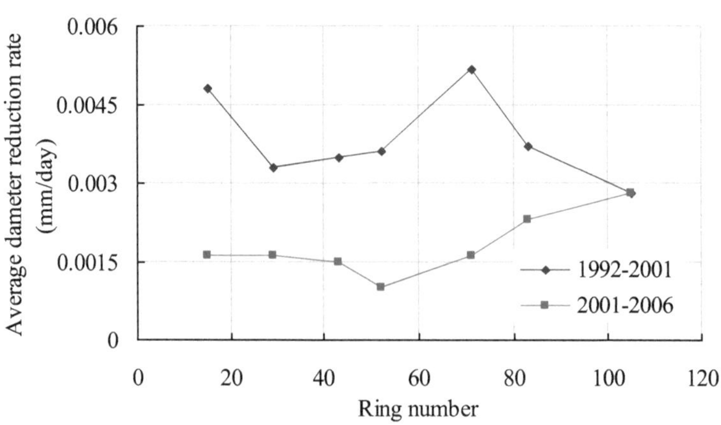

Fig. 4.11 In situ average diameter reduction rate

past years. The diameter reduction rates of the tunnel linings are in the order of 1‰ mm/day, which is in the same order as the quasi-steady axial creep rate of laboratory tests under low deviatoric stresses. However, the comparison is not so strict because the periods lasted for several months for each deviatoric stress level in the laboratory, which may not be long enough to reach the quasi-steady state of creep as in situ.

4.3 THM Creep Tests

4.3.1 Test Methodology and Procedures

The preparation of samples and SBCW, as well as test equipment, is described in Sect. 2.2.1. Table 4.3 gives the basic physical information of the test samples for the THM creep behavior study.

The test condition of the triaxial creep tests is shown in Table 4.4. The test procedures can be divided into five stages (Fig. 4.12):

Stage 1 (Path I → II): The re-saturation stage (see Sect. 2.2.1.4).

Stage 2 (Path II → III): Heating stage. The temperature of the triaxial cell was increased to 40 °C at an average rate of 0.3 °C/h.

Stage 3 (Path III → IV): Deviatoric stress loading stage. The axial load was increased to the requested value.

Stage 4 (Path IV → V → IV): Creep stage. The axial load was kept constant for about 5–7 months and underwent a heating–cooling cycle. A cyclic temperature process (40 → 60 → 80 → 60 → 40 °C) was regulated, which was similar to that expected in the real repository. Each heating sub-step or cooling sub-step was controlled at the heating rate of 0.30 °C/h and kept for at least 1 month. The creep test was conducted under drained conditions, which were facilitated by opening the drainage valves.

Table 4.3 Physical information for the samples of the creep tests

Sample	Density (g/cm^3)	Water content (%)	Dry density (g/cm^3)	Void ratio	Saturation degree (%)
L-THM-3.7-3	2.02	22.42	1.65	0.66	93.32
L-THM-3.7-4	2.02	22.42	1.65	0.66	93.32
L-THM-3.7-5	2.02	20.71	1.67	0.64	88.72
L-THM-3.7-6	2.02	20.71	1.67	0.64	88.72

Table 4.4 Test condition of THM-coupled drained triaxial creep tests

Sample	Confining pressure (MPa)	Back pressure (MPa)	Deviatoric stress (MPa)	Temperature (°C)
L-THM-3.7-3	3.7	2.2	0.5	40 → 60 → 80 → 60 → 40
L-THM-3.7-4	3.7	2.2	0.5	40 → 60 → 80 → 60 → 40
L-THM-3.7-5	3.7	2.2	1.0	40 → 60 → 80 → 60 → 40
L-THM-3.7-6	3.7	2.2	1.0	40 → 60 → 80 → 60 → 40

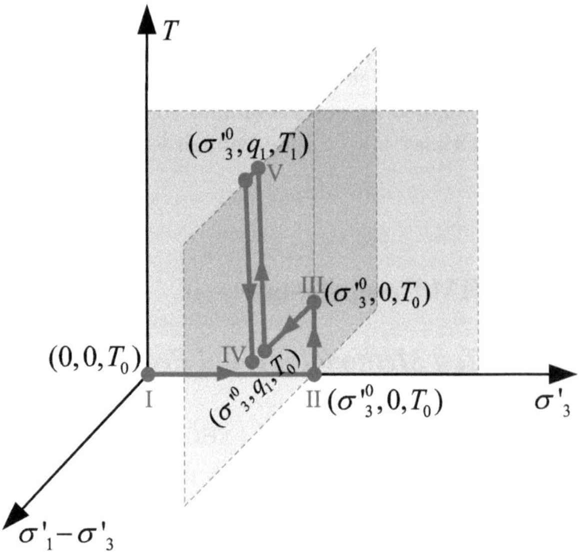

Fig. 4.12 Thermo-mechanical path of the THM-coupled creep test

4.3.2 Results of the THM Triaxial Creep Test

4.3.2.1 Temperature Effect on the Creep Behavior of Boom Clay

Test L-THM-4.7-1 was conducted under the effective confining pressure of 2.5 MPa. The deviatoric stress was kept at 1.5 MPa during the creep stage. Figure 4.13 shows the variation of axial strain and radial strain of test L-THM-4.7-1 during the creep stage. It is noted that some technical problems with the radial deformation transducer occurred during the creep stage, so the radial deformation was absent after the 1500th hour in the figures. Figure 4.14 shows the axial strain rate in the cyclic temperature process. From these two figures, it can be found that the axial strain had an instantaneous increment when the creep load was applied, and then the creep deformation occurred. The axial strain rate was relatively high in the initial creep stage but slowed with time and eventually reached the steady state. The axial creep strain rate at the steady state was on the order of 10^{-7} h^{-1} at 40 °C.

An increase in temperature induced a drastic acceleration of the creep rate. When the temperature increased from 40 to 60 °C, the creep rate increased to the order of 10^{-5} h^{-1}, and a new steady state was reached in about 20 days. The axial steady-state creep strain rate at 60 °C was 10^{-6} h^{-1}. When the temperature increased from 60 to 80 °C, the creep rate increased to the order of 10^{-5} h^{-1}. Thermal expansion was not observed in the heating stage. The creep rate increased so rapidly that the thermal expansion was concealed in the creep strain.

In contrast, however, a decrease in temperature induced an obvious thermal contraction in the axial direction for a short time. No obvious creep deformation was observed in the cooling phase. The calculated thermal expansion coefficient

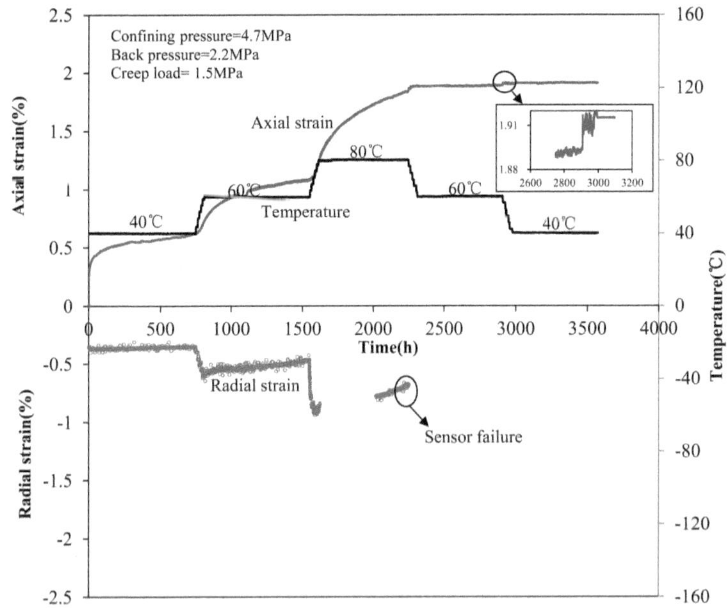

Fig. 4.13 Creep behavior of Boom Clay during a cyclic temperature process L-THM-4.7-1

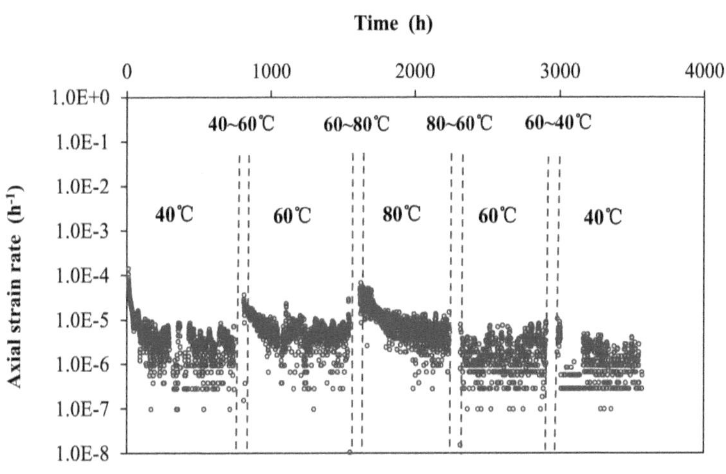

Fig. 4.14 Axial creep rate of Boom Clay with time and temperature (L-THM-4.7-1)

perpendicular to the bedding plane is about 1.5×10^{-5} K^{-1}. The creep behavior during the heating–cooling cycle found in our research is quite similar to the experimental results of Zhang et al. (2007), who investigated the HM-coupled creep behavior of CO_X claystone by conducting several uniaxial creep tests at various temperatures. In their tests, heating induced an increase in creep rate at the temperature range of 22–60 °C, while no creep deformation occurred during the cooling stage. The distinct creep behavior observed during the heating and cooling stages suggests that the creep behavior of clay rock is dependent on the thermal path followed. In the reheating stage, the creep behavior was activated again when the temperature was increased to the highest temperature in the previous creep stage.

4.3.2.2 Confining Pressure Effect on the Creep Behavior of Boom Clay

Test L-THM-3.7-1 was conducted under the effective confining pressure of 1.5 MPa. The deviatoric stress was kept at 1.5 MPa during the creep stage. Figure 4.15 shows the variation of axial strain and radial strain of test LTHM-3.7-1 during the creep stage. Figure 4.16 shows the axial strain rate in the cyclic temperature process. The radial deformation transducer failed soon after the temperature reached 80 °C. The evolution of the axial strain, radial strain, and creep strain rate during the creep stage is quite similar to test LTHM-4.7-1. However, because of the smaller confining pressure, the instantaneous strain caused by the axial load and the creep strain is much higher than in test LTHM-4.7-1. The axial strain at 80 °C was 8.08%, four times larger than that of test L-THM-4.7-1.

The creep strain rate was initially in the order of 10^{-4} h^{-1} when the sample was heated to 80 °C. When the temperature was kept for a month, the creep rate was still in the order of 10^{-5} h^{-1}. It can be concluded that, along with temperature, the creep rate also depends on the confining pressure.

No obvious creep deformation was observed in the cooling phase. A decrease in temperature induced an obvious thermal contraction in the axial direction for a short

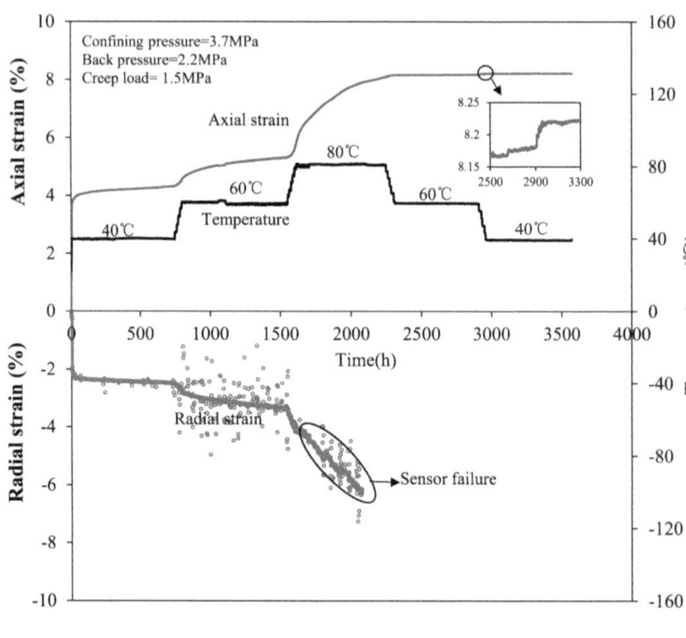

Fig. 4.15 Creep behavior of Boom Clay during a cyclic temperature process (L-THM-3.7-1)

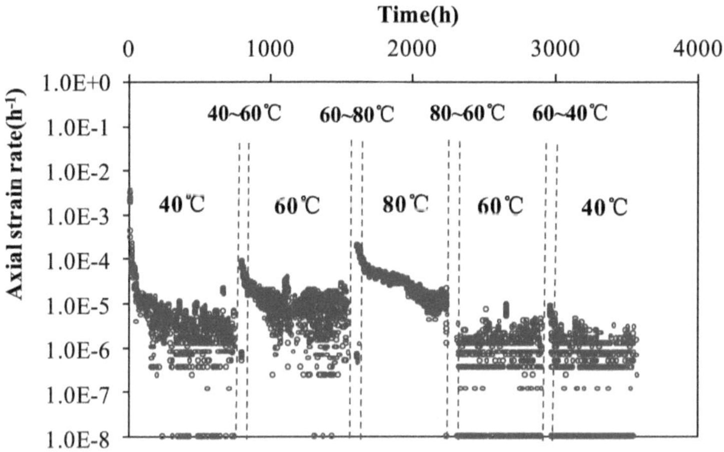

Fig. 4.16 Axial creep rate of Boom Clay with time and temperature (L-THM-3.7-1)

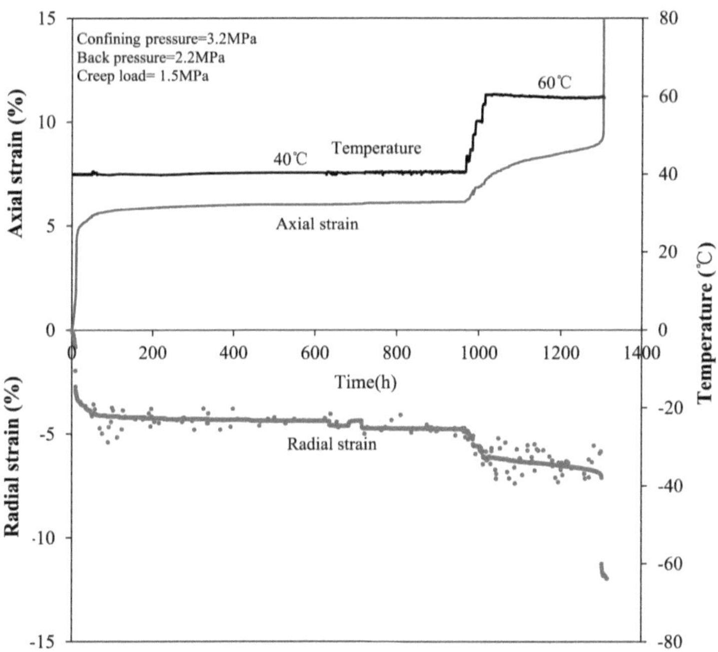

Fig. 4.17 Creep behavior of Boom Clay during a cyclic temperature process (L-THM-3.2-2)

time. The thermal expansion coefficient perpendicular to the bedding plane calculated from the results of test L-THM-3.7-1 is about 2.0×10^{-5} K^{-1}.

Test L-THM-3.2-2 was conducted under the effective confining pressure of 1.0 MPa. The deviatoric stress was kept at 1.5 MPa during the creep stage. Figure 4.17 shows the variation of axial strain and radial strain of test L-THM-3.2-2 during the creep stage. Figure 4.18 shows its axial strain rate in the cyclic

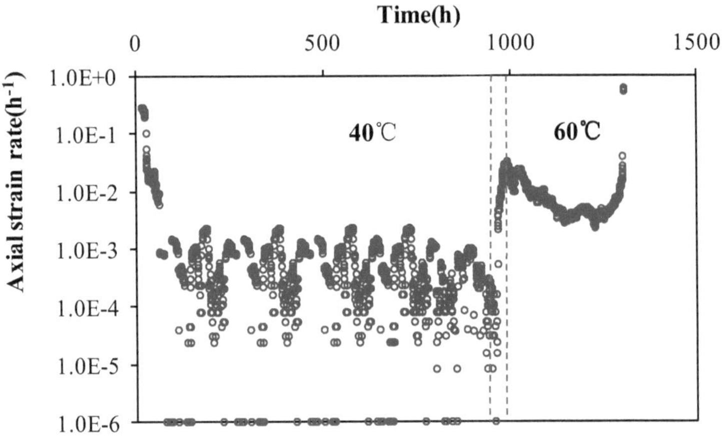

Fig. 4.18 Axial creep rate of Boom Clay with time and temperature (L-THM-3.2-2)

temperature process. Under the low effective confining pressure of 1.0 MPa, the sample failed due to the accumulation of creep strain after heating to 60 °C for about 10 days.

4.3.2.3 Deviatoric Stress Effect on the Creep Behavior of Boom Clay

Test L-THM-3.7-3 was conducted under the effective confining pressure of 1.5 MPa. The deviatoric stress was kept at 0.5 MPa during the creep stage. Figure 4.19 shows the variation of axial strain and radial strain of test L-THM-3.2-3 during the creep stage. Figure 4.20 shows its axial strain rate in the cyclic temperature process. The evolution of the axial strain, radial strain, and creep strain rate during the creep stage is quite similar to L-THM-3.7-1 and L-THM-4.7-1. However, because of the smaller deviatoric stress, the axial strain under 80 °C was 0.96%, only 1/8 of test L-THM-3.7-1. When the temperature was heated from 40 to 60 °C, the variation of creep strain rate was not significant. However, the creep rate increased to the order of 10^{-4} h^{-1} from 10^{-7} h^{-1} when the temperature increased from 60 to 80 °C.

The comparison of axial creep strain and strain rate between L-THM-3.7-1 and L-THM-3.7-3 is shown in Figs. 4.21 and 4.22. The axial creep strain rate of test L-THM-3.7-3 is about one order of magnitude lower than that of test L-THM-3.7-1 at the same temperature level.

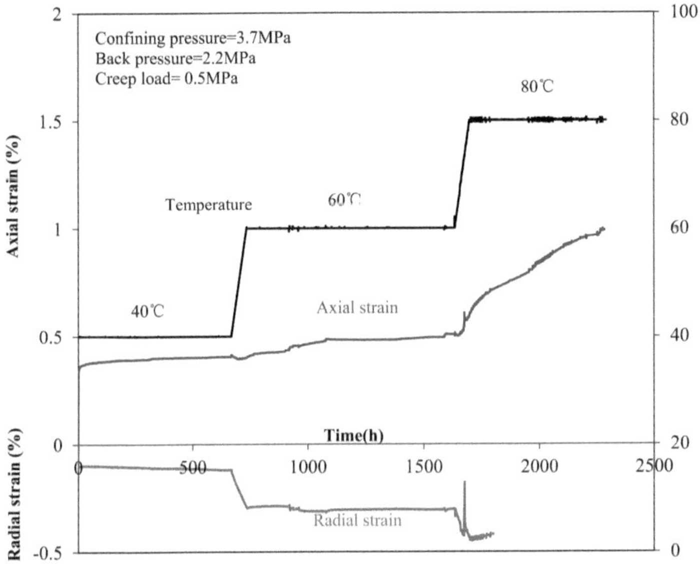

Fig. 4.19 Creep behavior of Boom Clay during a cyclic temperature process (L-THM-3.7-3)

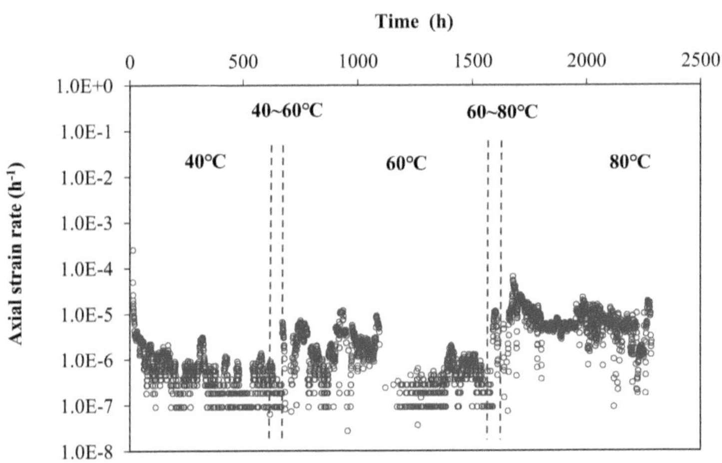

Fig. 4.20 Axial creep rate of Boom Clay with time and temperature (L-THM-3.7-1)

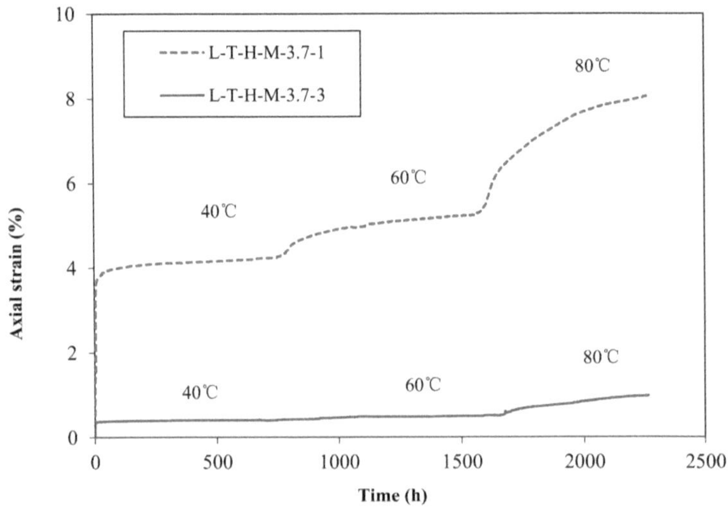

Fig. 4.21 The comparison of axial creep strain between L-THM-3.7-1 and L-THM-3.7-3

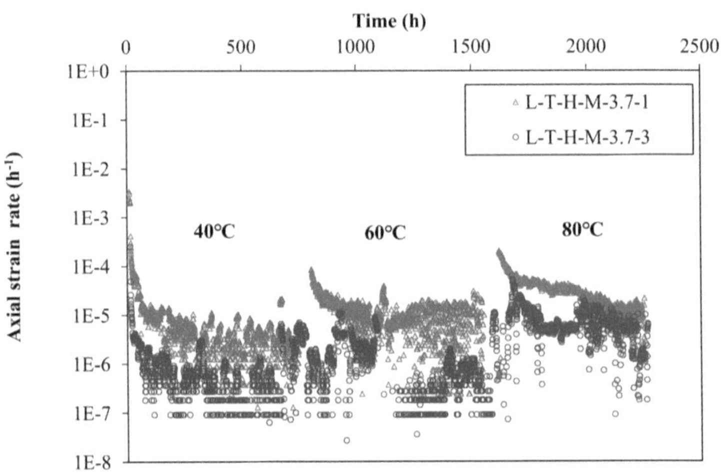

Fig. 4.22 The comparison of axial creep strain rate between L-THM-3.7-1 and L-THM-3.7-3

4.4 Mechanism of THM Creep Behavior of Boom Clay

4.4.1 Microstructure Change

4.4.1.1 Test Methodology

The environmental SEM technique was used to investigate the microstructure of Boom Clay before and after THM-coupled undrained triaxial creep tests. The environmental SEM techniques allow for imaging of fresh samples under undried and nonvacuum conditions, effectively preventing the development of desiccation cracks and unequal swelling of the clay particles.

The samples before THM-coupled undrained triaxial creep test were extracted from the sidewalls at tens of meters deep in the connecting gallery. This means that approximately "undisturbed" Boom Clay materials were used for SEM analysis to investigate their microstructural characteristics. Two samples (one parallel and another perpendicular to the bedding plane) have been investigated. To study the thermal impact on the microstructure of the Boom Clay, two further samples (one parallel and another perpendicular to the bedding plane) were taken from the sample after the completion of THM-coupled undrained triaxial creep tests.

4.4.1.2 Microstructure Change Before and After the THM Creep Tests

Figure 4.23 shows the microstructure of Boom Clay with a view parallel to the bedding. The apparent bedding was observed in SEM images of Boom Clay. This may be caused by the gravity of overburdening rock and frequent tectonic movements, along with the clay particles' preferred alignment parallel to the bedding plane. Dehandschutter et al. (2004, 2005) found similar results of fractured Boom Clay through SEM observations. Long, strip-shaped pores, measuring tens of microns in length and about 2 μm in width, have been observed between clay particles (Mag × 2000). These discontinuous, linearly arranged long, strip-shaped pores formed linear pore zones between adjacent beddings (Mag × 500). The (sub-horizontal) bedding is used as the reference for the transversely isotropic properties of Boom Clay in this book.

Figure 4.24 shows the microstructure of Boom Clay with a view perpendicular to the bedding. The clay particles contact by edge to face or face to face, bending into

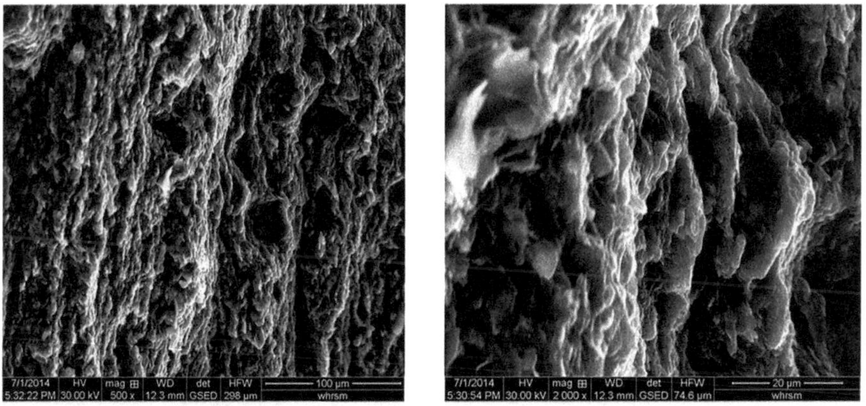

(a) Mag×500 (b) Mag×2000

Fig. 4.23 Microstructure of Boom Clay, view parallel to the bedding. (**a**) Mag × 500. (**b**) Mag × 2000 flaky structures (Mag × 500). The diameter of flaky structures is about tens of microns and is composed of smaller flaky particles. The diameter of flaky particles is only a few microns (Mag × 2000). These results are consistent with the previous results on the intact Boom Clay sample of Yu et al. (2010). However, significant differences can be seen between the SEM images with a view perpendicular and parallel to the bedding. Compared to the well-developed alignment of the clay particles in Fig. 4.23, the clay particles in Fig. 4.24 are more irregular. Moreover, no long, strip-shaped pores and linear pore zones have been observed between clay particles.

Figures 4.25 and 4.26 show the microstructure of Boom Clay samples after the completion of THM-coupled undrained triaxial creep test with views parallel and

(a) Mag×500 (b) Mag×2000

Fig. 4.24 Microstructure of Boom Clay, view perpendicular to the bedding. (**a**) Mag × 500; (**b**) Mag × 2000

(a) Mag×500 (b) Mag×2000

Fig. 4.25 Microstructure of Boom Clay after the completion of THM-coupled triaxial creep test, view parallel to the bedding. (**a**) Mag × 500; (**b**) Mag × 2000

(a) Mag×500 (b) Mag×2000

Fig. 4.26 Microstructure of Boom Clay after the completion of THM-coupled triaxial creep test, viewed perpendicular to the bedding. (**a**) Mag × 500; (**b**) Mag × 2000

perpendicular to the bedding, respectively. We can draw the following conclusion from the results: the anisotropic alignment of minerals was not disturbed by heating. Disintegration and rearrangement of clay particles reported by Almanza et al. (1995) were not observed.

The results of SEM tests highlight the transversely isotropic characteristics of intact Boom Clay. The apparent bedding observed can explain the anisotropy property of basic HM parameters of Boom Clay. For instance, these discontinuous, elongated-shaped pores highlight the higher horizontal permeability along the bedding plane.

4.4.2 Mechanism of THM Creep Behavior of Boom Clay

The axial strain had an instantaneous increment when the creep load was applied at the beginning of the tests. The axial strain rate was relatively high in the initial creep stage, and then the slope of the creep strain curve became smaller; the creep rate decreased rapidly with time and entered the stable creep stage from the transient creep stage. The creep strain curve is almost an inclined straight line, and the strain rate is constant.

This creep behavior of Boom Clay in the heating stage can be explained by the thermally induced reduction of viscosity and long-term strength. The mechanism of these thermal effects can be attributed to the microstructural weakening behavior. In addition, no significant thermal expansion was observed during the heating process, indicating that the creep deformation upon heating is mainly due to the thermally induced creep.

In contrast, however, decreasing the temperature induced an obvious cold contraction in the axial direction for a short time. The calculated thermal expansion coefficient perpendicular to the bedding plane is about 1.5×10^{-5} K^{-1}. That means cold contraction dominates the deformation behavior at the cooling stage of the THM-coupled creep tests. No obvious creep deformation was observed in the cooling phase. The creep behavior in the cooling phase suggests that the weakening of long-term strength caused by heating has no further evolution during cooling.

4.5 Conclusions

Triaxial creep tests were conducted at various temperatures (room temperature and high temperatures) and stress states to investigate the THM-coupled creep behavior and its mechanism in Boom Clay. The conclusions can be drawn as follows:

1. The HM creep tests (under room temperature) show that the creep behavior of Boom Clay became more and more significant as the deviatoric stress increased. The creep rate depends on deviatoric stresses, and for most of the loading steps (except the failed step) in the laboratory, the creep rate of Boom Clay reached a constant value after nearly 20 days. However, the in situ measurements show that the diameter reduction rate of Test Drift becomes stable approximately 7 years later.
2. The THM creep tests show that the creep rate (axial strain) increases with temperature elevation in the heating stage. This interesting phenomenon may be caused by the thermally induced reduction of viscosity and long-term strength of the clay skeleton. The mechanism of these thermal effects can be attributed to the weakening behavior. Under the stress paths applied in this study, cooling results in thermal compression (axial strain). No obvious creep appeared during the cooling phase, which suggests that the weakening of long-term strength caused by heating has no further evolution during cooling. Moreover, the creep rate also depends on the confining pressure and deviatoric stress.
3. The SEM tests performed before and after the THM creep tests show that the anisotropic alignment of minerals was not disturbed by heating.

Chapter 5
A Thermo-Elasto-Plastic Creep Damage Model for Boom Clay

5.1 Introduction

Establishing the appropriate models that can describe the THM-coupled effect on the creep properties of the host rock is essential for the long-term reliability of the HLW repository. This chapter first proposes an elasto-plastic creep damage model based on Perzyna's overstress theory to describe the creep behavior of Boom Clay at ambient temperature. The model was implemented in ABAQUS by the CREEP subroutine and USDFLD subroutine. Parameters are determined by back analysis based on the experimental studies in Sect. 4.2.2. Then, a thermo-elasto-plastic creep damage model was developed based on the proposed elasto-plastic creep damage model. The creep hardening and damage laws take into consideration thermal effects. The strain of clay was divided into three parts: the elastic, plastic, and creep components, and the first two parts were defined in Chap. 3. The model was also implemented in ABAQUS by the CREEP subroutine and USDFLD subroutine. Parameters are determined by back analysis based on the THM-coupled creep test results of Gong (2015). A comparison is made between the predictions of the presented developed model and the experimental data.

5.2 Elasto-Plastic Creep Damage Model for Boom Clay

5.2.1 Perzyna Overstress Model

Perzyna (1966, 1971) developed an elastic-viscoplastic model based on the overstress theory. In this model, the stress can increase beyond the viscoplastic yield surface, and the distance between yield surfaces and the stress is defined as the "overstress" (Fig. 5.1).

W. Chen et al., *Thermo-Hydro-Mechanical Coupling in Clay Rock*,
https://doi.org/10.1007/978-981-95-2562-1_5

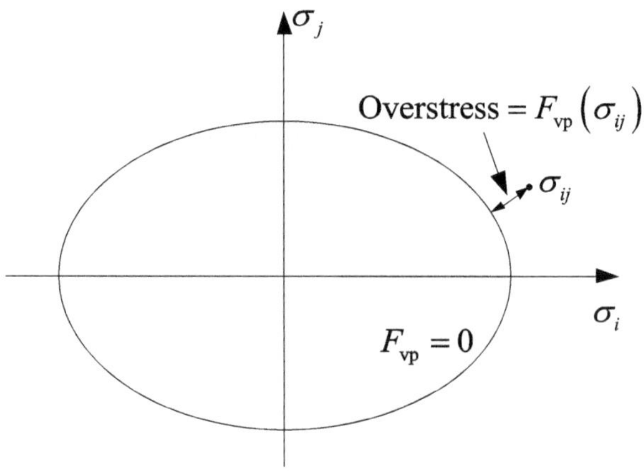

Fig. 5.1 Demonstration of Perzyna's overstress concept

Perzyna (1966) suggested that the strain is decomposed into two parts, i.e., the elastic part and the viscoplastic part.

$$\dot{\varepsilon} = \dot{\varepsilon}_e + \dot{\varepsilon}_{vp} \tag{5.1}$$

The viscoplastic strain rate intensity depends on the overstress. The value of the viscoplastic strain rate increases with the increasing overstress. It can be determined by

$$\dot{\varepsilon}_{vp} = \frac{1}{\mu}\left\langle \Psi\left(F_{vp}\right)\right\rangle \frac{\partial F_{vp}}{\partial \sigma} \tag{5.2}$$

in which μ is the material viscosity. Ψ is a scalar function of the overstress defined as

$$\Psi\left(F_{vp}\right) = \frac{F_{vp}}{F_0} \tag{5.3}$$

$\langle\rangle$ denotes a switch function (the Macaulay brackets), which can be defined as

$$\left\langle \Psi\left(F_{vp}\right)\right\rangle = \begin{cases} \Psi\left(F_{vp}\right) & ,\Psi\left(F_{vp}\right) > 0 \\ 0 & ,\Psi\left(F_{vp}\right) \le 0 \end{cases} \tag{5.4}$$

Overstress will decrease or increase over time due to the hardening/softening with the accumulation of viscoplastic strain. During the decelerating (/accelerating) creep process, the overstress decreases (/increases) with viscoplastic strain.

This model has become a standard basis for further development of viscoplastic models (František 2004) because it is simple but adequate in reproducing the creep behavior of a material under a complicated stress path.

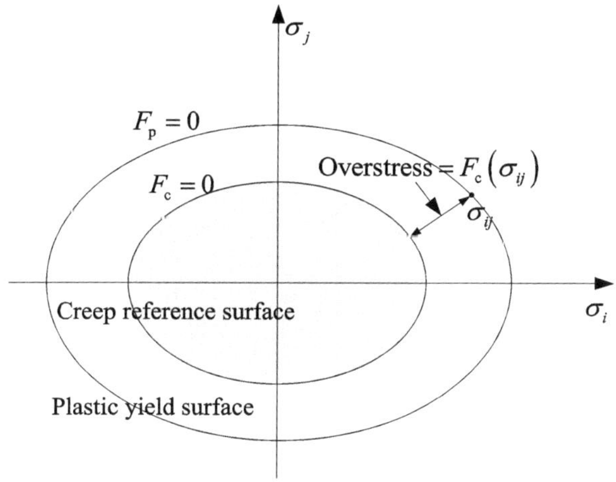

Fig. 5.2 Illustration of plastic yield surface and creep reference surface

Based on the experimental study on macro- and micro-rheology of clay, Dafalias (1982) suggested that the irreversible deformation of soil be divided into time-dependent and time-independent parts. Zhou et al. (2008, 2011) proposed two separate constitutive models for the description of instantaneous irreversible (plastic) deformation due to applied stress variation and time-dependent irreversible (creep) response due to material viscosity effect (Fig. 5.2.).

$$\dot{\varepsilon}_{ir} = \dot{\varepsilon}_p + \dot{\varepsilon}_{vp} \tag{5.5}$$

in which

$$\dot{\varepsilon}_p = \dot{\Lambda}\frac{\partial F_p}{\partial \sigma} \tag{5.6}$$

$$\dot{\varepsilon}_{vp} = \frac{1}{\mu}\langle \Psi(F_c)\rangle \frac{\partial F_c}{\partial \sigma} \tag{5.7}$$

Gong (2015) adopted this unified elastic–plastic–creep framework based on Perzyna's theory for simplicity and flexibility. In the present work, as an improvement and as a follow-up to the research program, the same framework was adopted to describe the creep behavior of Boom Clay.

5.2.2 *Creep Reference Surfaces and Equivalent Creep Surfaces*

Gong (2015) suggested that two different creep mechanisms are active in different loading regions: one is a cohesion mechanism, which follows the type of plasticity active in the shearing yield plasticity region, and the other is a consolidation

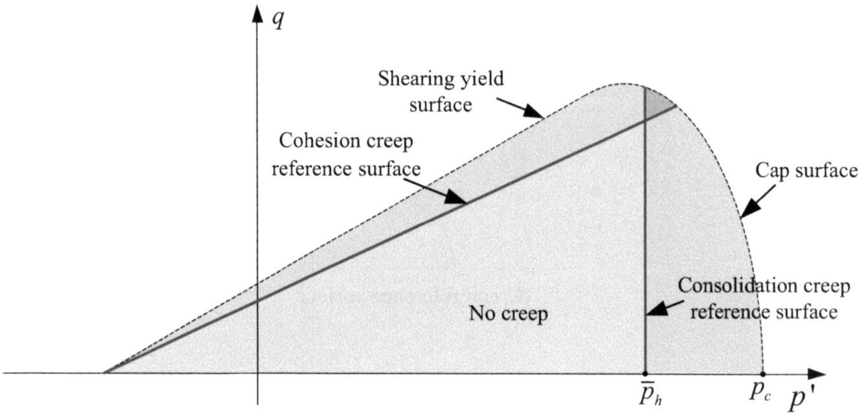

Fig. 5.3 Regions of activity of creep mechanisms

mechanism, which follows the type of plasticity active in the cap plasticity region. Figure 5.3. shows the region of applicability of the creep mechanisms in p–q space.

When considering the cohesion creep mechanism, the creep reference surface is given by:

$$F_{\mathrm{c}}^{\mathrm{s}} = q - \alpha_{\mathrm{c}} p - k_{\mathrm{c}} = 0 \tag{5.8}$$

in which α_{c} and k_{c} are two creep hardening parameters, $\alpha_{\mathrm{c}} \leq \alpha$, $k_{\mathrm{c}} \leq k$.

As consolidation creep is concerned, a simple expression of the creep reference surface is defined as

$$F_{\mathrm{c}}^{\mathrm{c}} = p - \bar{p}_{\mathrm{h}} = 0 \tag{5.9}$$

where $\bar{p}_{\mathrm{h}}$ is the creep reference consolidation pressure, which is also a creep hardening parameter.

It is assumed that there exists a creep isoclinic surface of stress points, which share the same "creep intensity" and are defined by an equivalent stress. This equivalent creep surface is parallel to the creep reference surface, as shown in Fig. 5.4. The cohesion equivalent creep reference surface can be converted to

$$F_{\mathrm{c}}^{\mathrm{s}} = \bar{\sigma}_{\mathrm{c}} - k_{\mathrm{c}} = 0 \tag{5.10}$$

where $\bar{\sigma}_{\mathrm{c}} = q - \alpha_{\mathrm{c}} p$ is the intersection point of the equivalent creep reference surface and the q axis.

The cohesion creep is dependent on the intersection point of the equivalent cohesion creep reference surface and the q axis, i.e., $\bar{\sigma}_{\mathrm{c}}$. If $\bar{\sigma}_{\mathrm{c}}$ is greater than a threshold value k_{c}, the cohesion creep is activated. Similarly, the activation of consolidation creep is dependent on whether the intersection point of the equivalent consolidation creep reference surface and the p-axis is above the threshold value $\bar{p}_{\mathrm{h}}$.

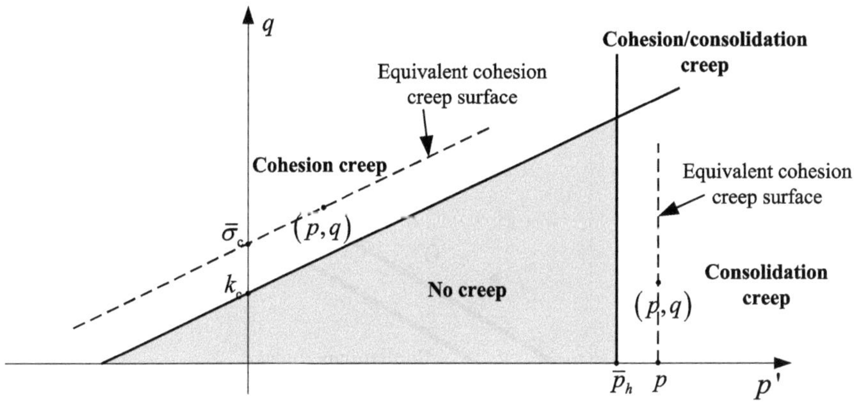

Fig. 5.4 Schematic diagram of the equivalent creep reference surfaces

Thus, the following four cases are considered in the model:

CASE 0: $p \le \overline{p}_h, \overline{\sigma}_c \le k_c$. No creep is activated.
CASE 1: $p > \overline{p}_h, \overline{\sigma}_c \le k_c$. Consolidation creep is activated.
CASE 2: $p \le \overline{p}_h, \overline{\sigma}_c > k_c$. Cohesion creep is activated.
CASE 3: $p > \overline{p}_h, \overline{\sigma}_c > k_c$. Both cohesion creep and consolidation creep are activated.

5.2.3 *Creep Flow Rules*

The creep strain produced by the cohesion mechanism is assumed to follow a potential that is a hyperbolic function (Fig. 5.5):

$$G_c^s = \sqrt{\left[\frac{0.1k}{(1-\alpha/3)}\alpha\right]^2 + q^2} - \alpha p \tag{5.11}$$

This continuous and smooth creep flow potential ensures that the flow direction is always uniquely defined.

The cohesive creep rate is intensity-dependent on the overstress and can be defined as

$$\dot{\varepsilon}_c^s = \frac{1}{\mu_s}\left\langle\frac{F_c^s}{f_0}\right\rangle^n \frac{\partial G_c^s}{\partial \sigma} = \frac{1}{\mu_s}\left\langle\frac{\overline{\sigma}_c - k_c}{f_0}\right\rangle^n \frac{\partial G_c^s}{\partial \sigma} \tag{5.12}$$

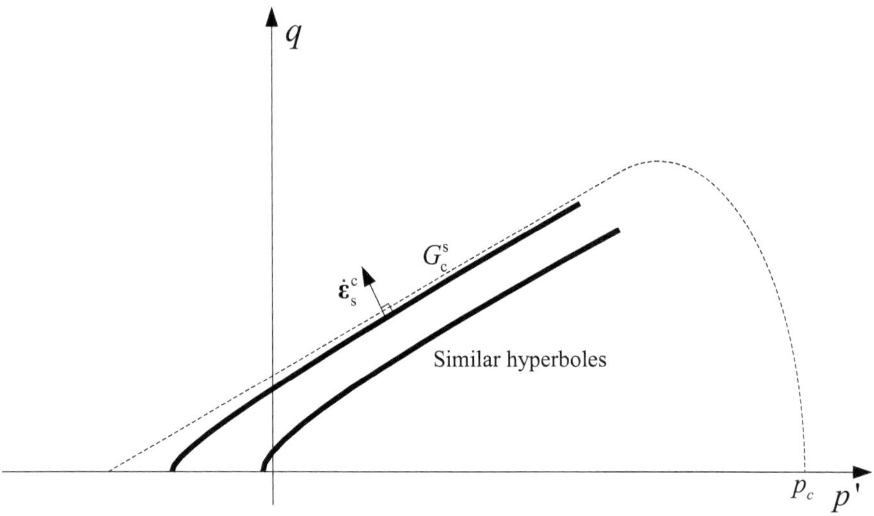

Fig. 5.5 Schematic diagram of the cohesive creep potential surface

The creep strain rate produced by the consolidation mechanism is assumed to follow a potential similar to that of the plastic strain rate in the cap yield surface (Fig. 5.6):

$$G_c^c = \sqrt{(p - p_a)^2 + (Rq)^2} \tag{5.13}$$

The consolidation creep rate can be defined as

$$\dot{\varepsilon}_c^c = \frac{1}{\mu_c}\left\langle \frac{F_c^c}{f_0} \right\rangle^n \frac{\partial G_c^c}{\partial \sigma} = \frac{1}{\mu_c}\left\langle \frac{p - \overline{p}_h}{f_0} \right\rangle^n \frac{\partial G_c^c}{\partial \sigma} \tag{5.14}$$

Thus, the creep rate of Boom Clay can be summarized as:

$$\dot{\varepsilon}_c = \begin{cases} 0, & \text{if } p \le \overline{p}_h, \overline{\sigma}_c \le k_c \\ \dfrac{1}{\mu_c}\left\langle \dfrac{p - \overline{p}_h}{f_0} \right\rangle^n \dfrac{\partial G_c^c}{\partial \sigma}, & \text{if } p > \overline{p}_h, \overline{\sigma}_c \le k_c \\ \dfrac{1}{\mu_s}\left\langle \dfrac{\overline{\sigma}_c - k_c}{f_0} \right\rangle^n \dfrac{\partial G_c^s}{\partial \sigma}, & \text{if } p \le \overline{p}_h, \overline{\sigma}_c > k_c \\ \dfrac{1}{\mu_c}\left\langle \dfrac{p - \overline{p}_h}{f_0} \right\rangle^n \dfrac{\partial G_c^c}{\partial \sigma} + \dfrac{1}{\mu_s}\left\langle \dfrac{\overline{\sigma}^{cr} - k_c}{f_0} \right\rangle^n \dfrac{\partial G_c^s}{\partial \sigma}, & \text{if } p > \overline{p}_h, \overline{\sigma}_c > k_c \end{cases} \tag{5.15}$$

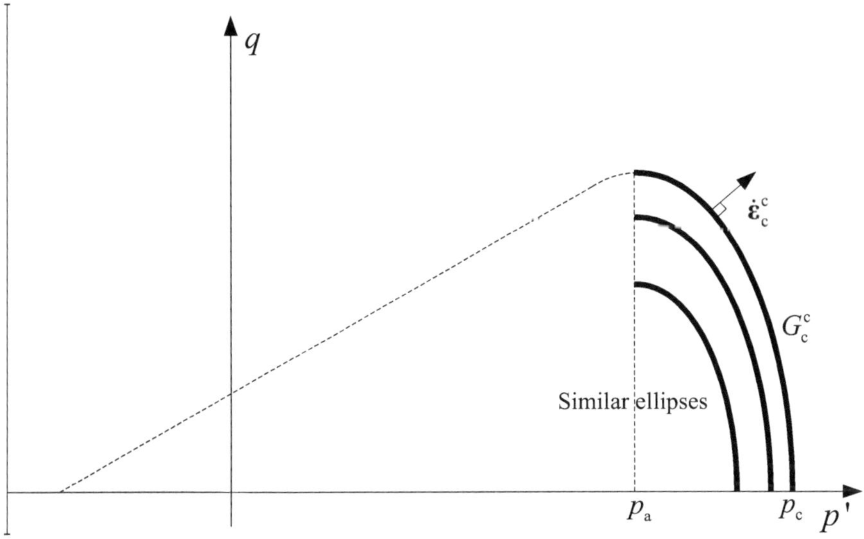

Fig. 5.6 Schematic diagram of the consolidation creep potential surface

5.2.4 Creep Hardening and Creep Damage

The creep hardening and creep damage evolution law were introduced to describe the variation of creep rate of Boom Clay during the creep stage. The forms of creep damage and creep hardening laws are the same as the proposed THMD coupled model in Chap. 3.

(1) Cohesion creep

The creep hardening evolution is similar to the plastic hardening function of the proposed THMD coupled model in Chap. 3. But the initial thresholds for creep hardening variables take lower values to ensure that the creep reference surface will not surpass the plastic yield surface. Based on the laboratory data in triaxial creep tests, the following nonlinear hyperbolic creep hardening parameter is proposed.

$$h_c = h_{c0} + \frac{b_c(1-h_{c0})\xi_c}{(1-h_{c0})+b_c\xi_c} \tag{5.16}$$

In which the equivalent deviatoric creep strain is defined by

$$\begin{cases} \dot{\xi}_{\mathrm{c}} = \sqrt{\frac{2}{3}\dot{\mathbf{e}}_{\mathrm{c}} : \dot{\mathbf{e}}_{\mathrm{c}}} \\ \mathbf{e}_{\mathrm{c}} = \boldsymbol{\varepsilon}_{\mathrm{c}} - \frac{1}{3}\mathrm{tr}\left(\boldsymbol{\varepsilon}_{\mathrm{c}}\right)\boldsymbol{\delta} \\ \xi_{\mathrm{c}} = \xi_{\mathrm{c}}\Big|_0 + \int_0^t \dot{\xi}_{\mathrm{c}} dt \end{cases} \tag{5.17}$$

where h_{c0} and b_{c1} are the model parameters.

A creep damage variable was introduced into instantaneous plastic hardening law to describe the shrinkage of the creep reference surface caused by the accumulation of the creep strain. The evolution of D_c is similar to the plastic damage evolution law of the proposed THMD coupled model in Chap. 3.

$$D_c = \bar{D}_c - \bar{D}_c \exp\left(-\beta_c \xi_c{}^d\right) \tag{5.18}$$

in which $\bar{D}_c$ is the upper limit value of D_c, and d and β_c are the model parameters.

It is assumed that ultimate values of internal creep hardening variables α_c and k_c are, respectively, equal to α_0 and k_0, which are defined in Chap. 3 as the ultimate values of plastic hardening variables. Thus, the evolution of α_c and k_c can be written as

$$\alpha_c = \alpha_0\left(1 - D_c\right)h_c \tag{5.19}$$

$$k_c = k_0\left(1 - D_c\right)h_c \tag{5.20}$$

(2) Consolidation creep

The evolution law of the threshold value $\bar{p}'_h$ considering the effects of irreversible volumetric strain on the consolidation creep behavior is assumed to be (Gong 2015):

$$\bar{p}'_h = \bar{p}_{h0} \exp\left(c_{ch}\varepsilon_c^v\right) \tag{5.21}$$

where

$$\varepsilon_c^v = \frac{1}{3}\mathrm{tr}\left(\varepsilon_c\right) \tag{5.22}$$

is the creep volumetric strain, $\overline{p}'_{h0}$ is the initial value of $\overline{p}'_h$, and c_{ch} is a model parameter.

5.2.5 *Secondary Development of the Proposed Model in ABAQUS*

Classical "creep" behavior of geomaterials that exhibit plasticity according to the capped Drucker-Prager plasticity model can be defined in ABAQUS (Abaqus 2006). The USDFLD subroutine is designed to update the model parameters and can be used to introduce the hardening or damage law of constitutive models. The CREEP subroutine allows for the definition of creep laws, which are combined with rate-independent plastic behavior in a coupled manner. By the combined calling of these two subroutines, the proposed creep damage model can be implemented on the platform of ABAQUS (Abaqus 2006).

The calling processes of USDFLD and CREEP subroutines are as follows:

Step 1. The main program of ABAQUS provides stress, total strain, total strain increment, time increment, and other state variables at the start of the increment.

Step 2. Update damage variables, Young's modulus, friction angle, cohesion, and creep hardening parameters based on the plastic/creep hardening law and the plastic/creep damage evolution law. Modify the model parameters by using the USDFLD subroutine.

Step 3. Determine whether the creep is activated. If the creep is activated, calculate the creep strain according to the proposed creep damage model.

Step 4. Calculate the elasto-plastic strain increment, $\Delta\varepsilon_{ep} = \Delta\varepsilon - \Delta\varepsilon_c$.

Step 5. Calculate the elastic trial stress. Perform loading/unloading check and update the elastic and plastic strain, stress, and consistent tangent stiffness with the constitutive integration algorithm.

Step 6. Return to the main program of ABAQUS. Perform equilibrium iteration. If the calculation converges, start the next increment. Otherwise, go to Step 1 with a smaller time increment until the calculation converges.

The flow chart of the subroutine is shown in Fig. 5.7.

The elastic modulus of 700 MPa and the Poisson ratio of 0.25 were taken in the developed CREEP subroutine.

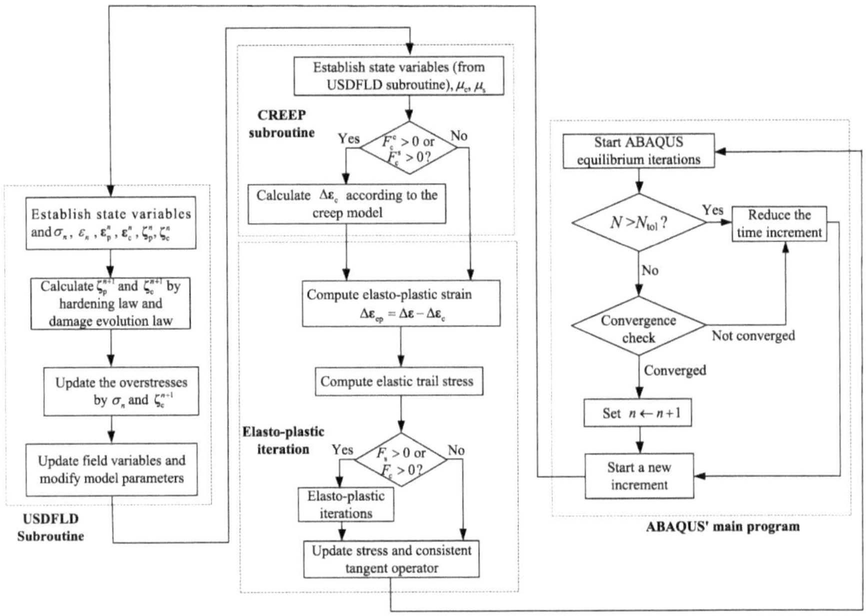

Fig. 5.7 Flow chart of CREEP and USDFLD user subroutines in the ABAQUS code

5.2.6 *Parameter Inversion and Simulation of the HM Coupled Creep Tests*

Similar to the method in Sect. 3.7.2, back analysis was used to determine the parameters of the creep damage model. Creep strain is the main test data that can be used in the back analysis. The creep strain error between numerical calculations and tests is used in the least-squares method. Back analysis needs to minimize the objective function, and the objective function of the back analysis is:

$$\varphi \rightarrow \min \tag{5.23}$$

The objective function based on the creep strain can be expressed as

$$\varphi\left(n, u_s, \beta_c, d, \bar{D}_c, h_{c0}, b_c, \cdots\right) = \sum_{k=1}^{l}\left(\varepsilon_{cij}^{\text{calk}} - \varepsilon_{cij}^{\text{testk}}\right)^2 \rightarrow \min \tag{5.24}$$

where $\varepsilon_{cij}^{\text{test}}$ are the test data corresponding to axial strains, $\varepsilon_{cij}^{\text{calk}}$ are the back analysis calculated value, and l is the number of test data used for back analysis.

Figure 5.8 shows the 3D finite element model for the Boom Clay sample with a dimension of Φ 38 × 76 mm. The bottom of the model is fixed in the axial direction, and then the bottom center of the model is fixed in both the axial and radial directions. The boundary conditions are the same as in the test.

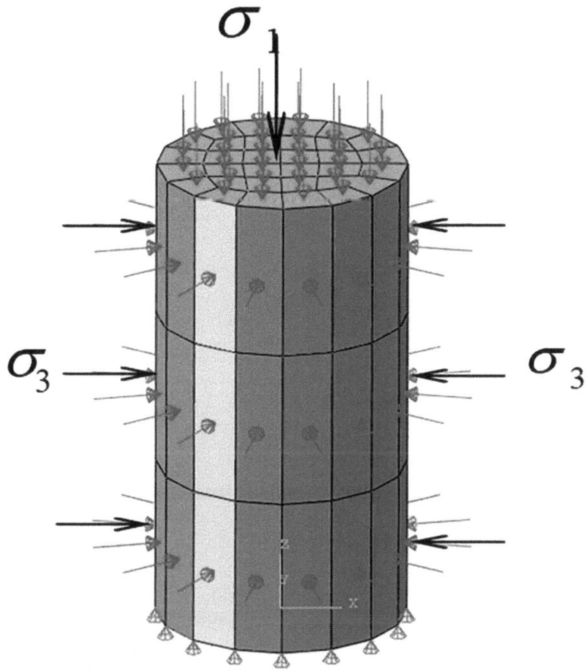

Fig. 5.8 Schematic diagram of the finite element model for test simulations

Table 5.1 Parameters of the elasto-plastic creep damage model for Boom Clay

Parameter	Value	Parameter	Value
n	5	f_0	1
μ_s	2×10^4	$\bar{D}_c$	0.5
μ_c	2×10^5	β_c	2500
h_{c0}	0.2	d	2.5
b_c	180	$\bar{p}_{h0}$/MPa	4.7
c_{ch}	62		

The calibration of parameters is done using the test data obtained by Sect. 4.2. Table 5.1 shows the back analysis results for the undetermined parameters. Figures 5.9 and 5.10 show the simulated and measured axial strain–time curve of tests TCP1 and TCP3. The creep behavior was reproduced by the elasto-plastic creep damage model. There is good agreement between experimental data, and numerical modeling validates the proposed elasto-plastic creep damage model.

Figures 5.11 and 5.12 show the evolution of cohesion overstress during the creep stage. The cohesion overstress reaches its maximum at the beginning of each creep stage and gradually decreases with time, remaining almost constant at the end of each creep stage.

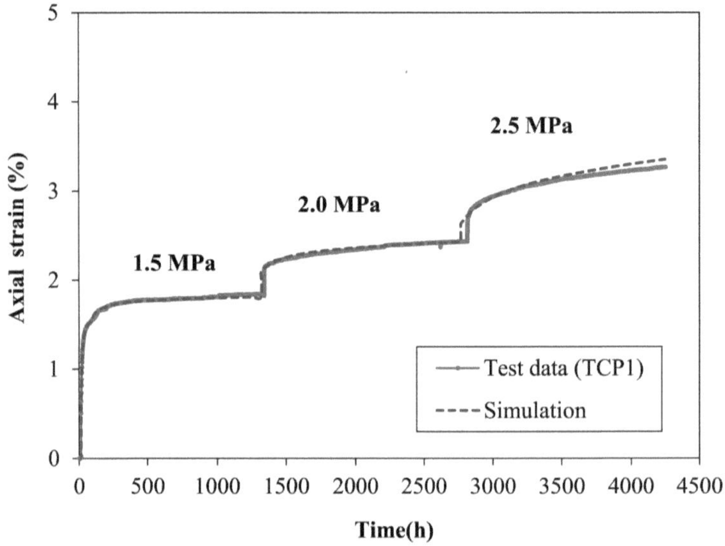

Fig. 5.9 Comparison between the simulated axial strain–time curve and the measured data (TCP-1)

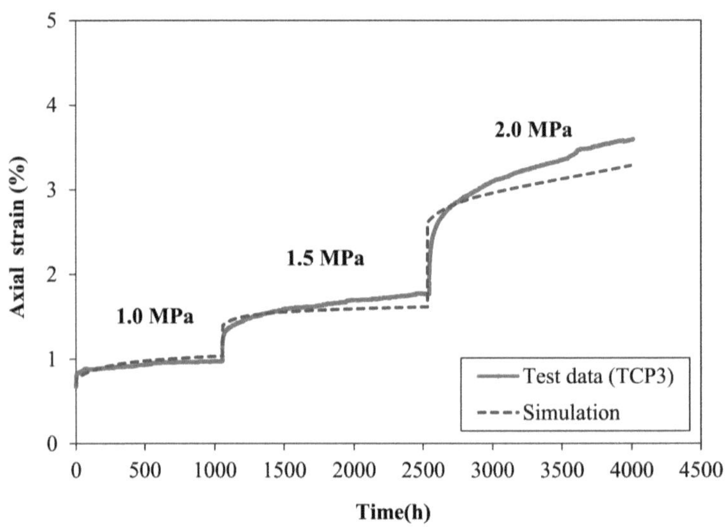

Fig. 5.10 Comparison between the simulated axial strain–time curve and the measured data (TCP3)

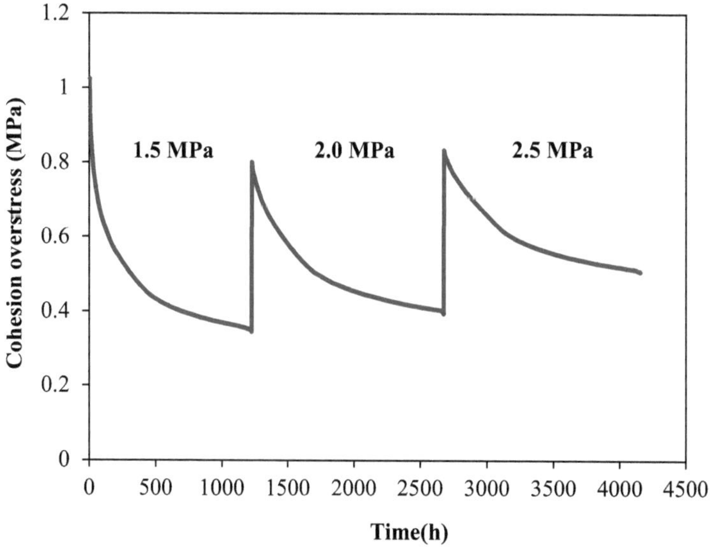

Fig. 5.11 Evolution of cohesion overstress during the creep stage (TCP-1)

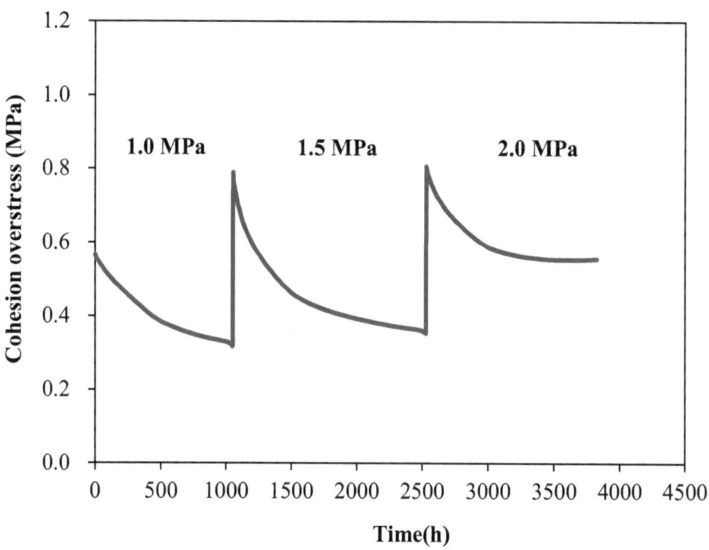

Fig. 5.12 Evolution of cohesion overstress during the creep stage (TCP3)

5.3 Thermo-elasto-Plastic Creep Damage Model for Boom Clay

5.3.1 Thermal Weakening Effect and Creep Hardening Law Considering Thermal Effects

The creep hardening law takes the same form as in Sect. 5.2.4. Further, the thermal effect on the creep behavior of Boom Clay was considered. Meanwhile, the thermal damage model was introduced as a further weakening effect on mechanical behavior due to temperature.

(1) Cohesion creep

Taking into consideration the thermal effects on creep hardening, the evolution law of h_c takes the following form:

$$h_{cT} = h_{c0} + \frac{b_c(1-h_{c0})(\xi_c - b_{cT}\Delta T)}{(1-h_{c0}) + b_c(\xi_c - b_{cT}\Delta T)} \tag{5.25}$$

where b_{cT} is a model parameter.

A thermal damage variable was introduced as a further weakening effect on mechanical behavior due to temperature. The evolution of D_{cT} is similar to the thermal damage evolution law of the proposed THMD coupled model in Chap. 3.

$$D_{cT} = 1 - \exp\left[-\beta_{cT}(T - T_0)\right] \tag{5.26}$$

in which

$$T = \begin{cases} T, & T > T_c \\ T_c, & T \le T_c \end{cases} \tag{5.27}$$

where T is the temperature in °C, T_0 is the initial temperature, T_c is the highest temperature in the geological history, and β_{cT} is a constant.

The THM-coupled creep tests (Sect. 4.3.2) show that heating weakens the long-term strength, and that the creep behavior is related to thermal history. Thus, the evolution of α_c and k_c under THM-coupled effect can be written as

$$\alpha_c = \alpha_0(1-D_c)(1-D_{cT})h_{cT} \tag{5.28}$$

$$k_c = k_0(1-D_c)(1-D_{cT})h_{cT} \tag{5.29}$$

(2) Consolidation creep

Based on a careful review of previous studies that comprehensively consider the effects of temperature and irreversible volumetric strain on the consolidation creep behavior, the evolution law of the threshold value $\overline{p}_h$ is assumed to be

$$\overline{p}_h = \overline{p}_{h0}\exp\left(c_{ch}\varepsilon_c^v\right)(1-D_{cT}) \tag{5.30}$$

5.3.2 Secondary Development of the Proposed Model in ABAQUS

The flow chart of the subroutines is shown in Fig. 5.13. The calling processes of USDFLD and CREEP subroutines are similar to Sect. 5.2.5, while the temperature increment is read as state variables, and the elasto-plastic strain increment is calculated considering thermal deformation: $\Delta\boldsymbol{\varepsilon}_{\rm ep} = \Delta\boldsymbol{\varepsilon} + \boldsymbol{\alpha}\Delta T - \Delta\boldsymbol{\varepsilon}_{\rm c}$

5.3.3 Parameter Determination by Back Analysis

Based on experimental studies, the back analysis has been performed to determine the thermal parameters of the proposed model. Creep strain is the main test data that can be used in the back analysis. For temperature and time, the creep strain error between numerical calculations and tests is used in the least squares method. The objective function of the back analysis is

$$\varphi \rightarrow \min \tag{5.31}$$

The objective function based on the creep strain can be expressed as

$$\varphi\left(\beta_{\rm cT}, b_{\rm cT}\right) = \sum_{k=1}^{l}\left(\varepsilon_{cij}^{\rm calk} - \varepsilon_{cij}^{\rm testk}\right)^2 \rightarrow \min \tag{5.32}$$

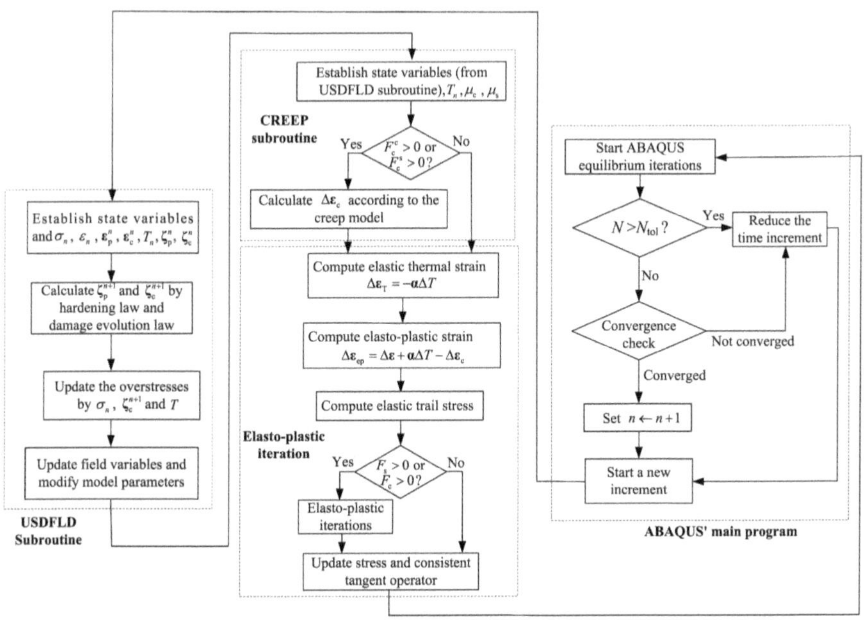

Fig. 5.13 Flow chart of CREEP and USDFLD user subroutines in the ABAQUS code

Table 5.2 Temperature-dependent parameters of the THM creep damage model

Parameter	Value
β_{cT}	0.0105
b_{cT}	0.000153

where $\varepsilon_{cij}^{\text{test}}$ are the test data corresponding to axial strains, $\varepsilon_{cij}^{\text{calk}}$ are the back analysis calculated value, and l is the number of test data used for back analysis.

The calibration of parameters is done using the test data. Table 5.2 shows the back analysis results for the undetermined parameters. Using parameters in Tables 5.1 and 5.2, the presented model is applied to simulate the creep tests under different stress and temperature conditions.

5.3.4 Simulation of the THM-Coupled Creep Behavior

Test L-THM-4.7-1 was a consolidated drained creep test conducted under the effective confining pressure of 2.5 MPa (i.e., the confining pressure of 4.7 MPa and pore water pressure 2.2 MPa). While test L-THM-3.7-1 is conducted under the effective confining pressure of 1.5 MPa (i.e., the confining pressure of 3.7 MPa and pore water pressure 2.2 MPa). A deviatoric stress of 1.5 MPa was applied at the beginning of the creep stage and was kept constant for about 5 months while undergoing a heating–cooling cycle. A cyclic temperature process (40 → 60 → 80 → 60 → 40 °C) was regulated to imitate the temperature changes of the host rock during the HLW disposal process. Each heating or cooling sub-step was controlled at the heating rate of 0.30 °C/h and kept for about 1 month. The creep test was conducted under drained conditions, which was done by opening the drainage valves.

Figures 5.14 and 5.15 show the simulated and measured strain–time curves of the test L-THM-4.7-1 and L-THM-3.7-1, respectively. The THM-coupled creep behavior was also obtained by numerical simulation with the proposed creep damage model. In the heating stage, the proposed model can reproduce the acceleration of the creep rate induced by increasing temperature well. The good agreement between experimental data and numerical modeling validated the applicability of the proposed model.

Figures 5.16 and 5.17 show the evolution of cohesion overstress during the creep stage of the tests L-THM-4.7-1 and L-THM-3.7-1, respectively. Similar to the HM coupled creep tests, the cohesion overstress reaches its maximum at the beginning of each heating stage. The cohesion overstress increased drastically at the beginning of each heating stage, then decreased slowly with time, and was almost constant at the end of each heating stage. In the cooling stage, however, the cohesion overstress decreased with temperature and then remained almost constant throughout the cooling stage, which indicates that the creep rate decreased to a negligible value.

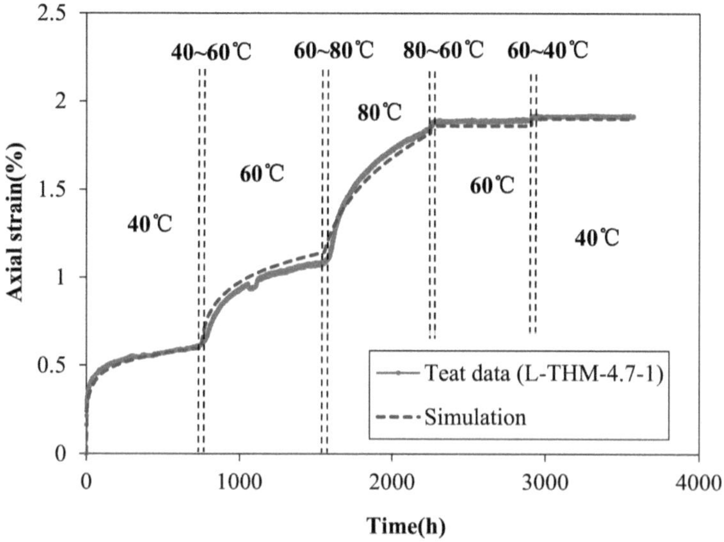

Fig. 5.14 Comparison between the simulated axial strain–time curve and the measured data (L-THM-4.7-1)

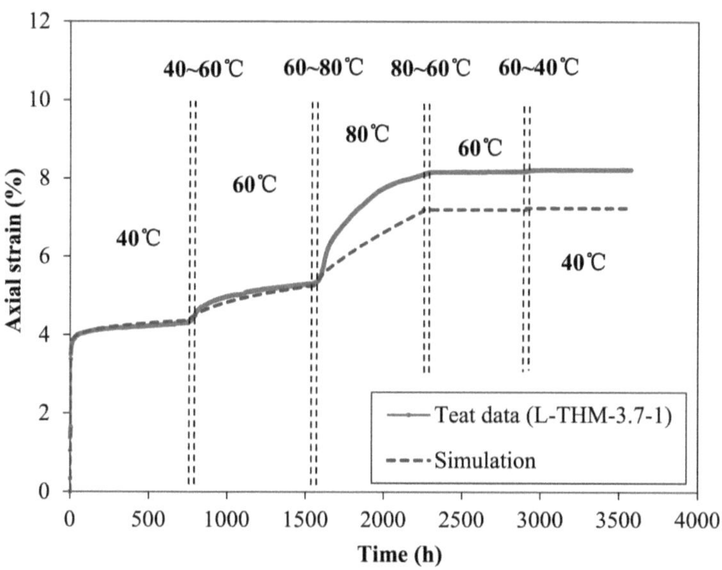

Fig. 5.15 Comparison between the simulated axial strain–time curve and the measured data (L-THM-3.7-1)

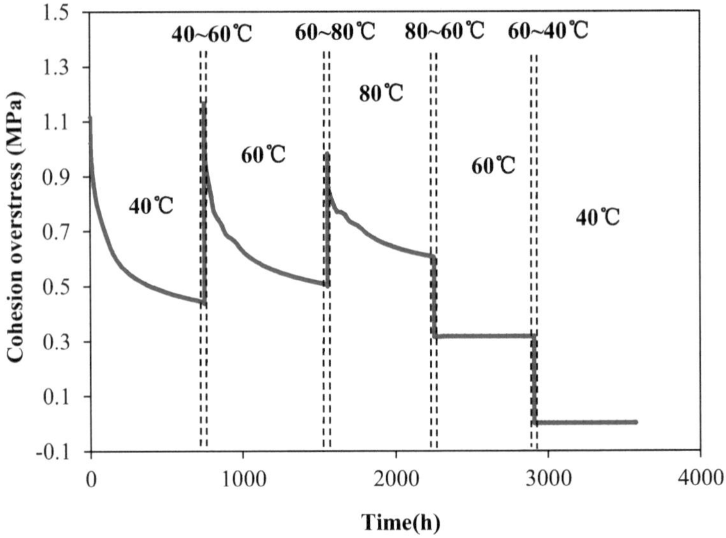

Fig. 5.16 Evolution of cohesion overstress during the creep stage (L-THM-4.7-1)

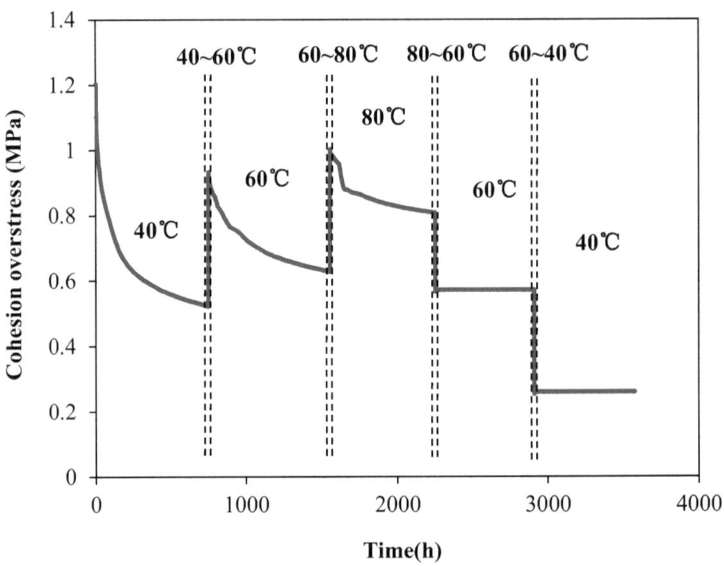

Fig. 5.17 Evolution of cohesion overstress during the creep stage (L-THM-3.7-1)

5.4 Conclusions

An elasto-plastic creep damage constitutive model is proposed first based on the overstress theory of Perzyna. This model is proven to be useful in describing the HM-coupled creep behavior of Boom Clay. Then, a thermo-elasto-plastic creep damage model was developed based on the proposed elasto-plastic creep damage model. The creep hardening law takes the same form as the elasto-plastic creep damage model. The temperature effect on the creep hardening behavior was considered. Furthermore, the thermal damage model was introduced as a further weakening effect on the mechanical behavior due to temperature. This model is an extension of the elasto-plastic damage model developed in Chap. 3. The model was implemented in ABAQUS by the CREEP subroutine and USDFLD subroutine. Parameters were obtained by the Nelder–Mead back analysis method based on both the HM and THM creep tests of Boom Clay. The good agreement between simulated results and test data validated the applicability of the proposed model.

Chapter 6
HM Perturbation During the Connecting Gallery Construction (CLIPEX)

6.1 Introduction

The formation of an excavation damaged zone (EDZ) is an inherent consequence of geologic repository construction, where induced fractures significantly alter the host rock's hydraulic properties. Meanwhile, excavation induces also the pore pressure change, in the host rock due to the Hydro-Mechancial coupling and creates a hydraulic disturbed zone (HdZ). Extensive research efforts have been dedicated to characterizing the hydromechanical (HM) behavior of EDZ and HdZ, focusing on three critical aspects: self-sealing capacity (Bastiaens et al. 2007; Monfared et al. 2012; Gonzalez-Blanco et al. 2024), time-dependent mechanical evolution (Blumling et al. 2007; Rutqvist et al. 2009), and fracture network development (Bossart et al. 2004; Armand et al. 2014). Underground research laboratories (URLs)—such as the Meuse/Haute-Marne URL in France, HADES URL in Belgium, and Mont Terri URL in Switzerland—have conducted systematic in situ experiments to quantify the HM responses of EDZ under realistic repository conditions (Bastiaens et al. 2003; Delay et al. 2007; Bernier et al. 2007; Yu et al. 2024b).

In this chapter, we employ a three-dimensional numerical model, developed through integration of prior theoretical studies and their numerical implementations, to systematically interpret the in situ measurements acquired during the CLIPEX project.

W. Chen et al., *Thermo-Hydro-Mechanical Coupling in Clay Rock*,
https://doi.org/10.1007/978-981-95-2562-1_6

6.2 CLIPEX Project

6.2.1 Project Overview

Among the extensive in situ experiments conducted at HADES, the CLIPEX project provides an invaluable platform for advancing our understanding of the hydromechanical (HM) behavior of the host rock clay. The European CLIPEX instrumentation program enabled high-fidelity characterization of the instantaneous HM responses exhibited by the clay during the excavation of the connecting gallery (Bernier et al. 2002). Systematic documentation of the excavated wall through continuous monitoring allowed detailed characterization of fracture patterns, a critical aspect, given that fracturing directly governs the HM response of the host rock.

The primary objective of the CLIPEX project was to evaluate the EDZ and HdZ around a gallery analogous in dimensions to a proposed radioactive waste disposal tunnel. Conducted within the framework of the URL extension in the deep Boom Clay formation at Mol, Belgium, this initiative involved constructing a secondary shaft and excavating a 90-meter-long connecting gallery linking the new shaft to the existing facility (Fig. 6.1). The CLIPEX research program systematically investigated the excavation-induced mechanical and hydrological responses through three key phases: (a) In situ characterization: Deployment of pressuremeters, dilatometers, and hydrofracturing tests to quantify natural stresses, permeability, deformability, and rock mass strength in the excavation-affected zone. (b) Predictive validation: Blind prediction exercises followed by rigorous comparison with field measurements to assess model accuracy. (c) Real-time monitoring: Installation of sensor networks at the advancing face of the Test Drift to capture transient hydromechanical (HM) behavior of the Boom Clay.

A critical outcome of this comprehensive dataset reveals two unprecedented findings: (a) The development of an EDZ coupled with intense HM coupling effects that classical elasto-plastic constitutive models fail to fully capture; (b) significant

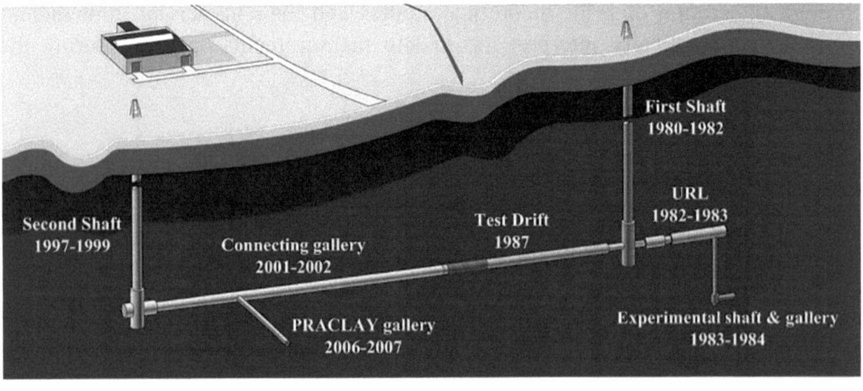

Fig. 6.1 Layout of the HADES underground research facility (Chen et al. 2022a, b)

discrepancies in pore water pressure evolution patterns observed in the far field compared to theoretical predictions. These results underscore the complexity of near-field HM interactions and highlight limitations in current predictive frameworks for deep geological repositories.

6.2.2 Instrumentation Installation

The instrumentation layout comprises four zones (Fig. 6.2) with dual boreholes per zone (A1–D1 for displacement and A2–D2 for total stress/pore water monitoring):

6.2.2.1 Zone Configuration

- Zone A: Axial region near connecting gallery axis
- Zone B: Vertical plane above gallery with 4° upward inclination
- Zone C: Same vertical plane as B but 11° upward inclination
- Zone D: Radial divergence from gallery axis

6.2.2.2 Borehole Instrumentation

Displacement Monitoring
- A1: Axial extensometer for gallery axis displacement
- B1: Inclinometer (4° upward) tracking radial movements.
- C1: Inclinometer (11° upward) monitoring steeper radial displacements
- D1: Deflectometer measuring horizontal displacement vectors

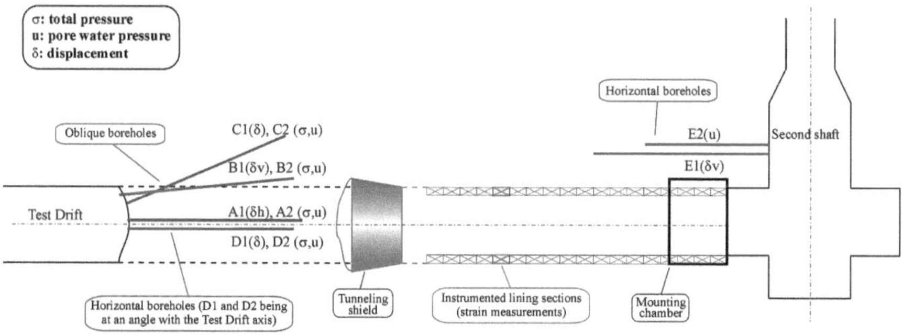

Fig. 6.2 Instrumentation on and around the second shaft and test drift (Bastiaens et al. 2003)

Total Stress/Pore Water Monitoring

- All paired boreholes (A2–D2) feature:
- 6-filter pressure transmitters (8-filter in C2) for pore water pressure
- Total stress cells for geomechanical response characterization

This configuration enables systematic comparison between displacement evolution (A1–D1) and coupled hydromechanical responses (A2–D2).

6.3 In Situ Measurements and Their Interpretation

6.3.1 Pore Water Pressure Measured from Test Drift

The temporal evolution of pore pressures, quantified as a function of distance from the advancing face to each sensor, is presented in Fig. 6.3. Key findings from these measurements reveal:

The temporal evolution of pore pressure recorded by all sensors exhibits a consistent pattern: exhibiting ~5.5 bar differential between WA1 and WA6 at the beginning and progressive increase culminating in a peak value, followed by a decline as the excavation front approaches. Sensors positioned at greater distances (C2) demonstrate post-decline pressure recovery. The observed pressure increase phase corresponds to the undrained contracting plastic behavior of the clay mass, while the subsequent decline is attributed to dual mechanisms: decompression-induced fracturing around the excavation front and localized fluid flow re-equilibration driven by pore pressure gradients. The pressure recovery phenomenon at greater distances (C2) emerges from two concurrent processes: recompression of clay induced by equilibrium between gallery wall convergence and support system reaction forces and hydromechanical coupling restoring pore pressure equilibrium through massif decompression-triggered fluid migration.

Notably, all sensors exhibit simultaneous activation immediately following excavation initiation, as evidenced by Fig. 6.4. The monitored pore pressure began to rise when the excavation front remained 60 m distant, demonstrating a hydromechanical disturbance zone extending at least 60 m ahead of the advancing face. This phenomenon underscores the rapid propagation of hydraulic perturbations during tunneling operations.

6.3.2 Displacements Measured from Test Drift

The displacement evolutions are redrawn as a function of the front position, as shown in Fig. 6.5. Note that the extensometer in borehole A1 measures effectively the axial displacements along the axis of the gallery.

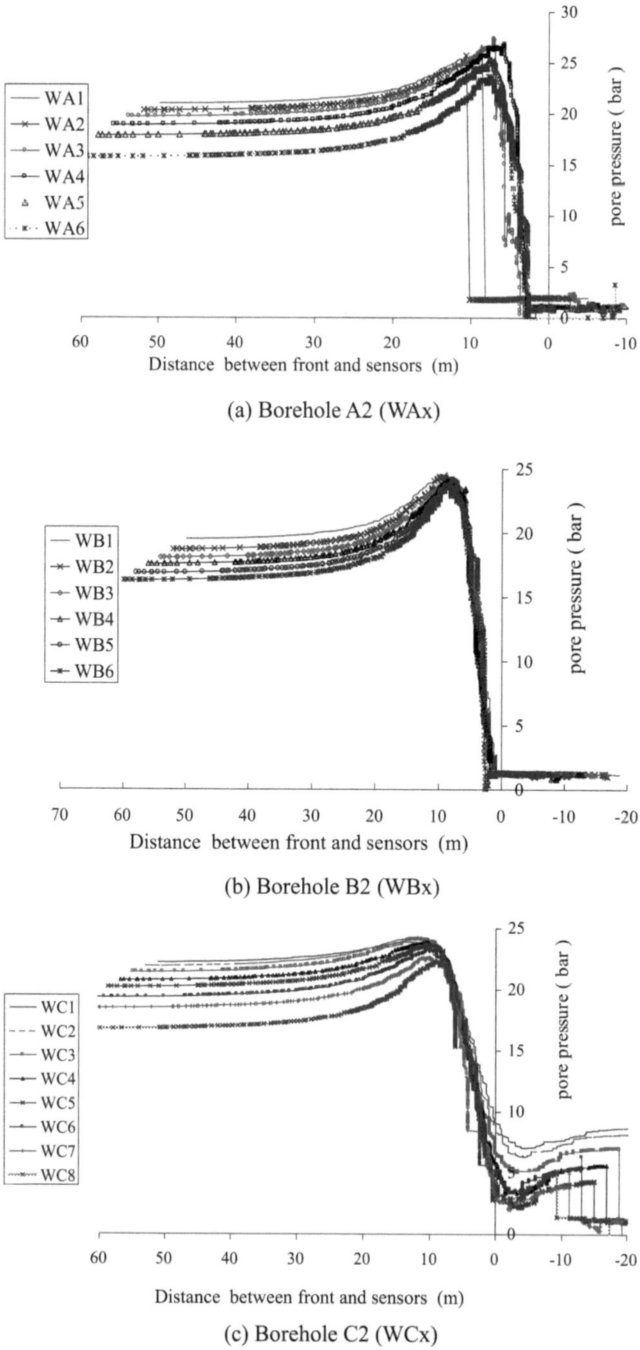

Fig. 6.3 Pore pressure evolution as a function of the excavation front position (Jan et al. 2022). (**a**) Borehole A2 (WAx); (**b**) borehole B2 (WBx); (**c**) borehole C2 (WCx)

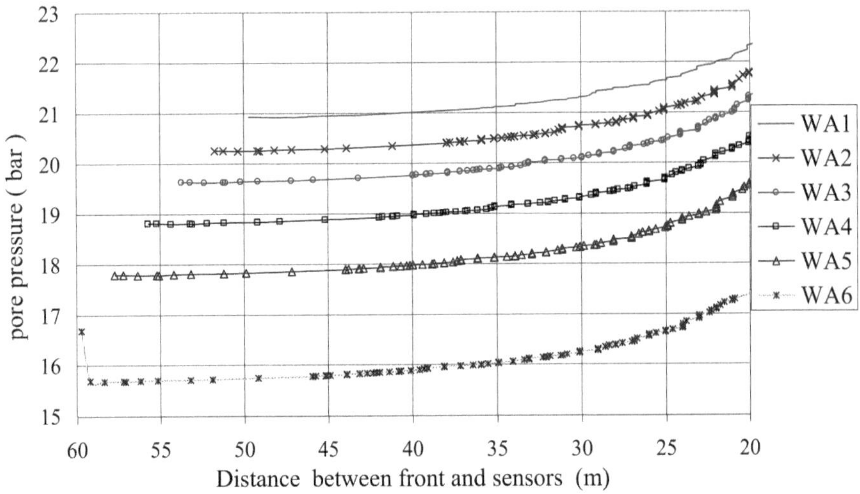

Fig. 6.4 Far-field reactions to excavation: Borehole A2 (Wax) (after Jan et al. 2022)

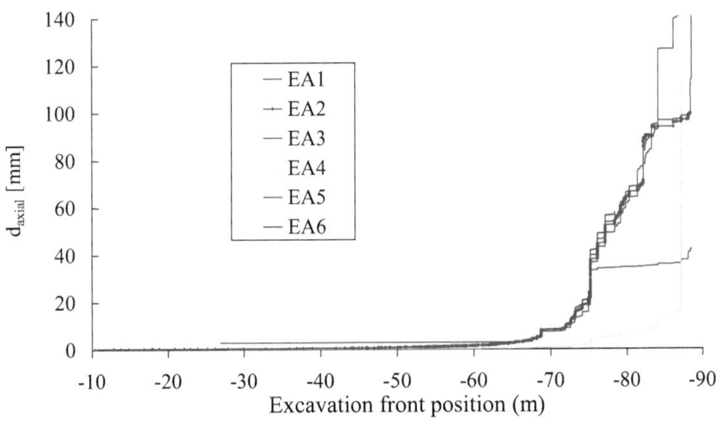

Fig. 6.5 Displacement evolution during C.G. excavation (from Test Drift): Borehole A1 (after Jan et al. 2022)

Although the interpretation of these measurements presents challenges, valuable insights have been obtained:

6.3.2.1 Axial Displacement Along Gallery Axis

Recordings corroborate the hypothesis that all sensors were rigidly mounted together, with signal transmission consolidated to a single point: the disk installed approximately 5 m ahead of the T.D. front. All sensors provided instantaneous recordings that terminated when the T.D. front reached approximately 5 m from its target position. Final axial displacement measurements ranged between 120 and 140 mm (see global view in Fig. 6.5).

6.3.2.2 Lining Installation Effects

Detailed analysis reveals distinct saw-toothed patterns in sensor outputs, clearly reflecting the lining installation effects. Notably, while extensometer A1 failed to capture data at all expected positions, it nevertheless yielded reliable measurements at a critical axis-aligned point. This data proves particularly valuable for characterizing the hydromechanical behavior of Boom Clay.

6.4 Numerical Simulation

The blind prediction study utilized actual excavation parameters, including advancement rate, sequencing, excavated cross-section dimensions, and lining geometry.

6.4.1 Hypotheses

- Three-dimensional modeling incorporated vertical planar symmetry relative to the gallery axis.
- Anisotropic stress distribution was explicitly considered.
- Influence from the second shaft was included in the analysis.
- A perfect bonding interface is assumed between the clay mass and lining (contact mechanics not explicitly modeled)
- Undrained excavation conditions were maintained throughout the simulation.

6.4.2 Geometry

The Cartesian coordinate system (O, X, Y, Z) implemented in the numerical model is illustrated in Fig. 6.6. The vertical *Z*-axis points upward, with the *X*-axis aligned with the tunnel axis. The tunnel axis is positioned at a depth of 222.9 m below ground level, where vertical planar symmetry relative to its longitudinal plane has been incorporated into the model configuration.

The numerical model spans 235 m along the *X*-axis (comprising 67 m for the Test Drift and 90 m for the connecting gallery), 50 m in the *Y*-direction, and 100 m in the *Z*-direction. The computational domain discretization comprises 75,875 finite

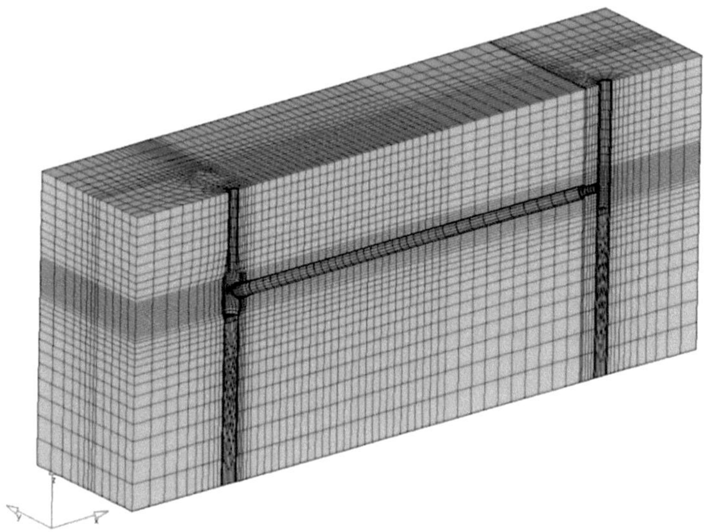

Fig. 6.6 General view of the finite element model

elements and 69,216 nodal points. Figure 6.6 presents the comprehensive geometry of the numerical simulation model

The Test Drift excavation radius measures 2.55 m with a 2.40 m external liner radius, resulting in a 15 cm annular gap between the host rock and support structure. The composite liner system maintains 0.60 m structural thickness throughout. Figure 6.7 demonstrates the geometric configuration of the Test Drift excavation.

The numerical modeling of the connecting gallery specifies an excavation radius of 2.405 m with 0.40 m lining ring thickness, as detailed in the geometric configuration shown in Fig. 6.8.

Figure 6.9 provides a detailed visualization of the Test Drift-connecting gallery junction, demonstrating the simulated sequential excavation process between these critical underground structures.

6.4.3 Numerical Modeling Procedures

The coupled hydromechanical simulations were conducted within a saturated porous medium framework, following the construction chronology through six distinct phases:

Stage I: First Shaft Excavation (1980–1982)

The first shaft construction was modeled using ABAQUS's simplified methodology: instantaneous element deactivation in target excavation zones, followed by incremental installation of concrete lining elements for structural reinforcement.

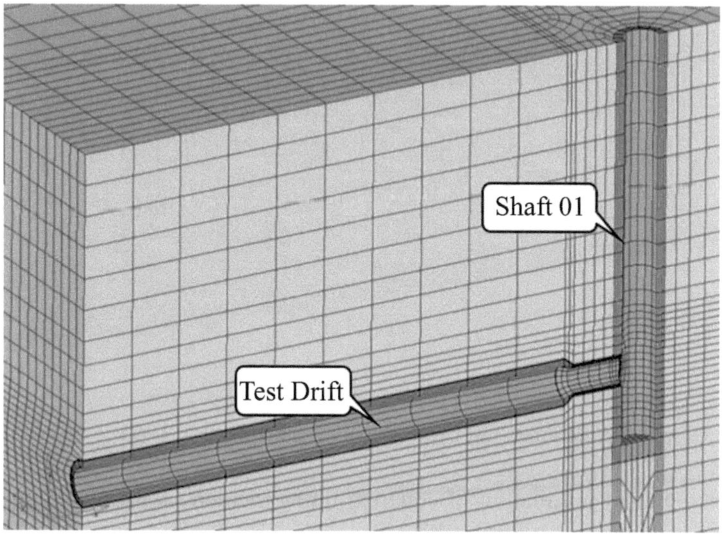

Fig. 6.7 Geometry of the Test Drift

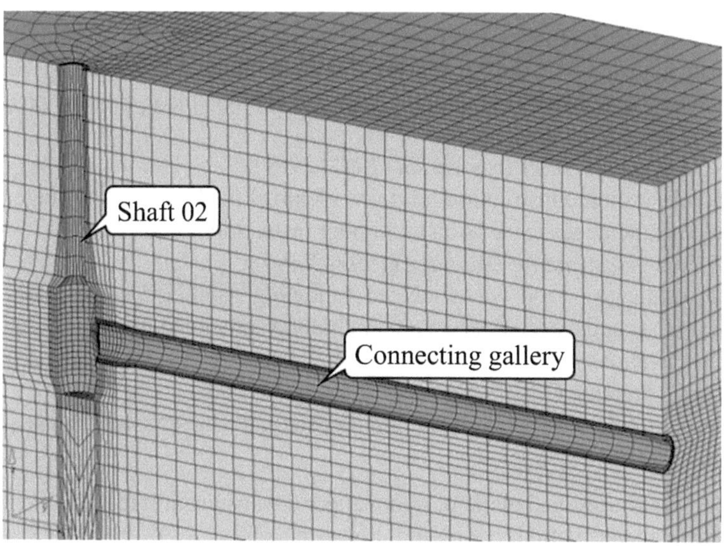

Fig. 6.8 Geometry of the connecting gallery

Stage II: Test Drift Excavation (1987)

Notably, comprehensive instrumentation deployment at the Test Drift face necessitated explicit simulation of clay mass unloading history. This involved ABAQUS's simplified approach, combining:

- Immediate void creation through element deactivation
- Linear pressure relaxation across tunnel boundaries to achieve static equilibrium

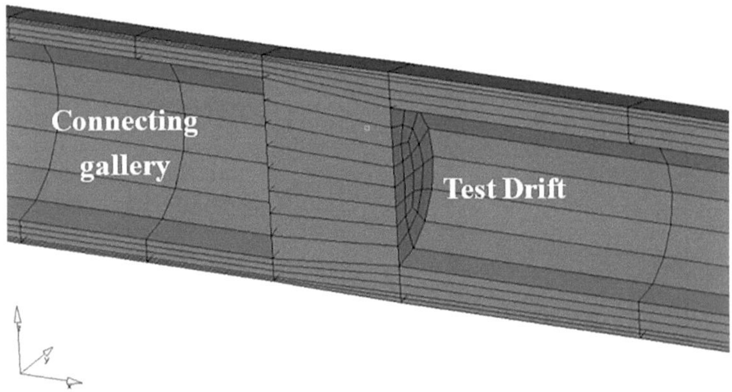

Fig. 6.9 Geometry of the joint (Test Drift-connecting gallery)

- Incremental installation of concrete lining elements with material property assignments

The problem is considered an undrained condition as the hydraulic conductivity of Boom Clay is very low.

Stage III: Post-Drift Drainage Period (1987–1997)

A 10-year hydromechanical response period was simulated following Test Drift completion and before secondary shaft excavation, incorporating progressive water drainage mechanisms.

Stage IV: Secondary Shaft Excavation (1997–1999)

The secondary shaft construction followed identical methodology to Stage I, replicating instantaneous excavation and lining installation sequences.

Stage V: Post-Construction Drainage (1999–2001)

A 2-year drainage period was analyzed following combined completion of Test Drift and secondary shaft structures, modeling integrated drainage effects from both excavations.

Stage VI: Connecting Gallery Construction (2001–2003)

Shield-supported gallery excavation implemented realistic construction sequencing through:

- Full cross-sectional element removal per advancement interval
- Simultaneous boundary pressure linear relaxation
- Progressive installation of 90 m-long concrete lining segments

This incremental approach maintained synchronization between excavation progression and structural support installation. All excavations assumed undrained conditions at 3 m/day advancement rates. Shield protection was numerically represented through convergence constraints, while lining behavior was modeled using elastic constitutive relationships.

6.4.4 *Boundary Conditions*

6.4.4.1 Initial Stress Field Calibration

The model has been initialized to the in situ stress field before excavations according to the following:

- $\sigma_v = \rho g\text{h}$, vertical stress ($\sigma_v \cong 4.5$ Mpa at the gallery axis depth)
- $\sigma_x = \lambda_x \sigma_v$, horizontal stress in the direction parallel to the tunnel axis
- $\sigma_y = \lambda_y \sigma_v$, horizontal stress in the direction normal to the tunnel axis

- Pore water pressure was prescribed based on $u_w = \gamma_w h$, establishing initial stress states and fluid pressure distributions within the model domain. The upper boundary condition imposes vertical stress equivalent to the overburden pressure from the host rock strata.

6.4.4.2 Boundary Constraint Specifications

- Vertical displacement constraints were applied at the basal boundary (Z-direction).
- Full normal displacement constraints were enforced on all four lateral boundaries.

6.4.5 *Corresponding Parameters Used for the Simulations*

1. *Mechanical properties*
   For the parameters required for the Mohr-Coulomb model of Boom Clay, see Sect. 2.2.
2. *Hydraulic properties*
   All partners have done numerical simulations under the hydromechanical coupling effect and regarded the clay rock as saturated material. The main hydraulic parameters required for the simulations are given in Table 6.1.
3. *Lining properties*
   The linings were modeled using the mechanical elements by most of the partners and depicted by an elastic constitutive law with the following parameters (Table 6.2).

Table 6.1 Hydraulic properties of Boom Clay

Horizontal permeability K_h (m/s)	Vertical permeability K_v (m/s)	Porosity
4.32×10^{-12}	2.87×10^{-12}	0.39

Table 6.2 Properties of lining materials

	Young's modulus (GPa)	Poisson's ratio	Thickness (cm)	Excavation radius (m)	Saturated permeability (m/s)	Gap (cm)
Connecting gallery	35	0.25	40	2.445	3.6×10^{-11}	4.5
Test drift	25	0.25	60	2.55	3.0×10^{-11}	15

6.5 Comparison Between Field Measurements and Numerical Results

6.5.1 Pore Pressure

Figure 6.10 compares measured and simulated pore pressures from three piezometers (Boreholes A2–C2) versus connecting gallery excavation progress. The numerical predictions demonstrate reasonable accuracy in replicating field observations.

6.5.1.1 Accurate Reproduction of In Situ Stress Gradients

The CLIPEX measurements confirm significant preexcavation hydraulic gradients at all boreholes. Numerical simulations incorporating elastoplastic constitutive models and permeability evolution hypotheses successfully replicated these gradients through hydromechanical coupled analysis of Test Drift excavation effect and 14-year postexcavation water flow modeling. Good agreement with measurements is shown in the near-face Borehole A2. However, predicted gradients slightly overestimate field values discovered in Distal Boreholes B2/C2.

6.5.1.2 Consistent Evolution Patterns

Both measurement datasets and simulations exhibit characteristic pressure evolution patterns characterized by initial minor increases (0.2–2.5 bar) followed by rapid declines coinciding with excavation advancement. This phenomenon corresponds to soil skeleton dilation following excavation-induced stress redistribution (decompression effect). The numerical model reproduces this general behavior, showing progressive pressure buildup reaching peak values before decline. While sensors in boreholes B2 display increasing trends, these are less pronounced compared to the primary measurement locations.

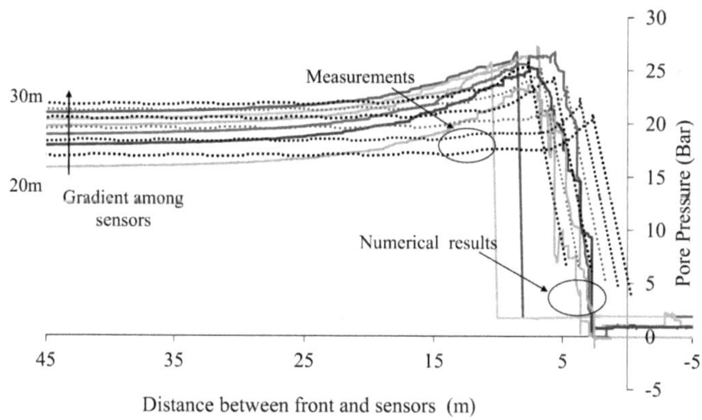

(a) Borehole A2

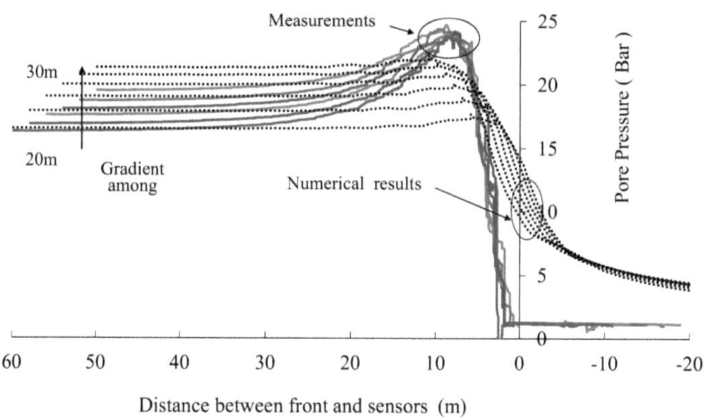

(b) Borehole B2

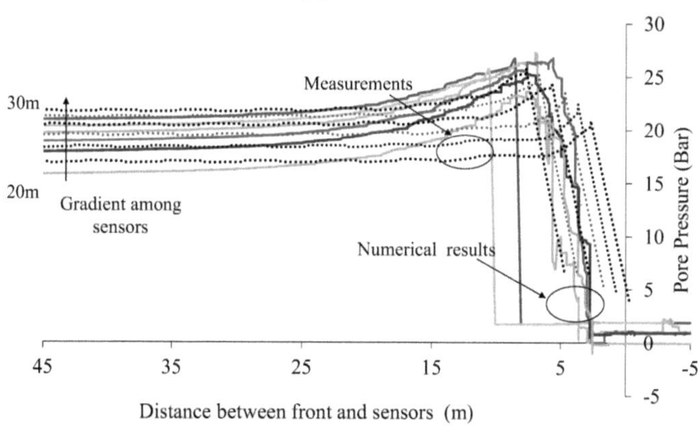

(c) Borehole C2

Fig. 6.10 Pore pressure comparisons between in situ measurements and predictions. (**a**) Borehole A2; (**b**) borehole B2; (**c**) borehole C2

6.5.1.3 Peak Value Comparison

Numerical predictions exhibit comparable peak magnitudes (maximum 3.5 bar vs. measured 8 bar, Fig. 6.10) but demonstrate delayed pressure peak occurrence and significantly reduced absolute pressure increments compared to CLIPEX field measurements.

6.5.1.4 Hydraulic Disturbed Zone (HdZ) Analysis

The numerical model underpredicts regional-scale hydraulic disturbance, with simulated HdZ extent at least twofold smaller than in situ observations. This discrepancy highlights challenges in capturing far-field hydraulic responses through current modeling frameworks.

6.5.1.5 Tunnel Face Pressure Dynamics (Borehole A2)

Predicting pressure attenuation near the excavation face is achieved through element depletion modeling, which eliminates pore pressure data in depleted zones (Fig. 6.10a). This corresponds with physical processes of material removal during gallery advancement.

6.5.1.6 Regional Pressure Attenuation (Boreholes B2/C2)

While numerical simulations replicate general pressure decay trends, measured pore pressure fluctuations exhibit substantially greater amplitude and spatial propagation. The model captures primary pressure evolution patterns but underrepresents dynamic response characteristics in distal regions.

6.5.1.7 Permeability Evolution

The permeability distribution hypothesis successfully predicts substantial permeability enhancement (Figs. 6.11 and 6.12) within damaged zones, demonstrating a strong correlation between hydraulic conductivity evolution and pressure redistribution patterns.

6.5.1.8 Excess Pore Pressure Distribution at R55 Section

Figure 6.13 illustrates the excavation-induced excess pore pressure distribution around the R55 section. The simulation reveals a pronounced pressure buildup along the horizontal axis of the connecting gallery (R55), with the maximum pressure occurring approximately 3.6 m from the tunnel wall in the horizontal direction.

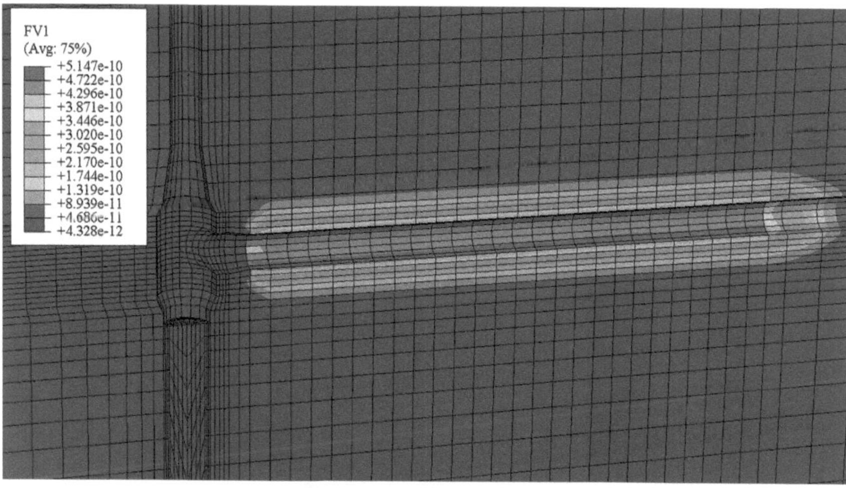

(a) Distribution of horizontal permeability K_h

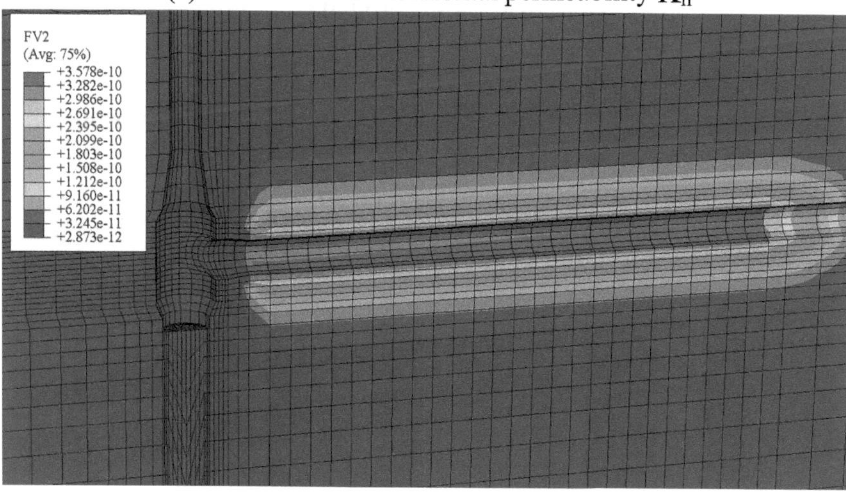

(b) Distribution of vertical permeability K_v

Fig. 6.11 Permeability variation in the excavation zone. (**a**) Distribution of horizontal permeability K_h; (**b**) distribution of vertical permeability K_v

6.5.2 Displacement

Figure 6.14 compares field-measured and numerically simulated axial displacements at borehole A1 relative to connecting gallery excavation progression. Key findings reveal:

Field measurements demonstrate parallel displacement trends across sensors (Fig. 6.14), indicating uniform axial strain distribution. The numerical model

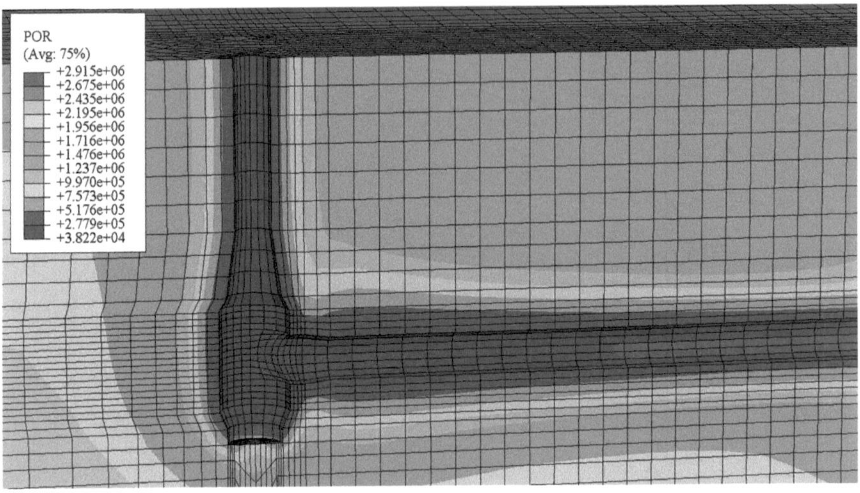

(a) Pore pressure distribution around connecting gallery

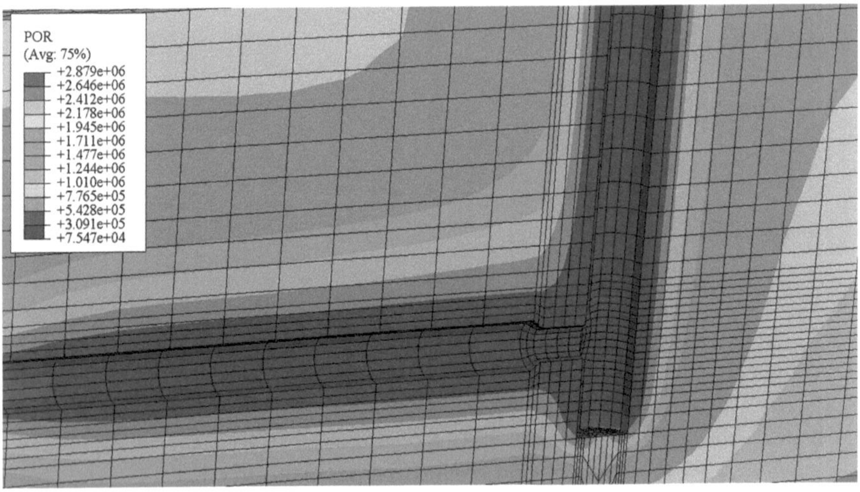

(b) Pore pressure distribution around test shaft

Fig. 6.12 Pore pressure distribution around gallery excavation. (**a**) Pore pressure distribution around connecting gallery; (**b**) pore pressure distribution around test shaft

successfully replicates this characteristic pattern, achieving close agreement in final displacement magnitudes (140–160 mm). The model exhibits delayed displacement initiation compared to field observations—numerical responses emerge only when the excavation front advances within 20 m of the monitoring point. In contrast, field sensors detect immediate displacement changes at excavation commencement. This temporal lag results in underestimation of the EDZ extent, where the simulated EDZ measures approximately half the size of field-observed disturbances.

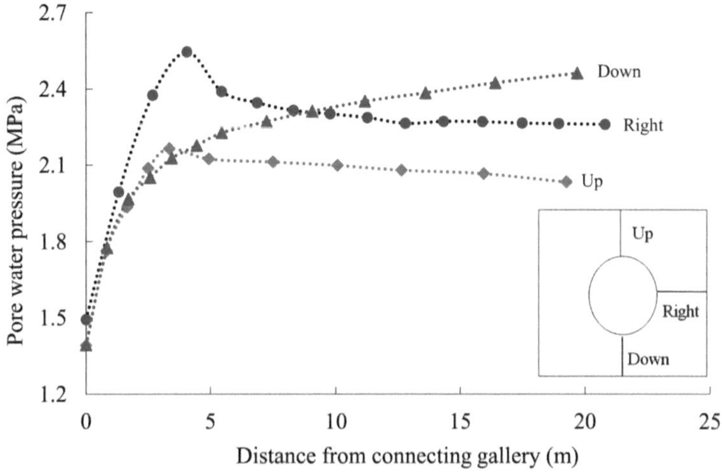

Fig. 6.13 Excess pore water pressure distribution around R55

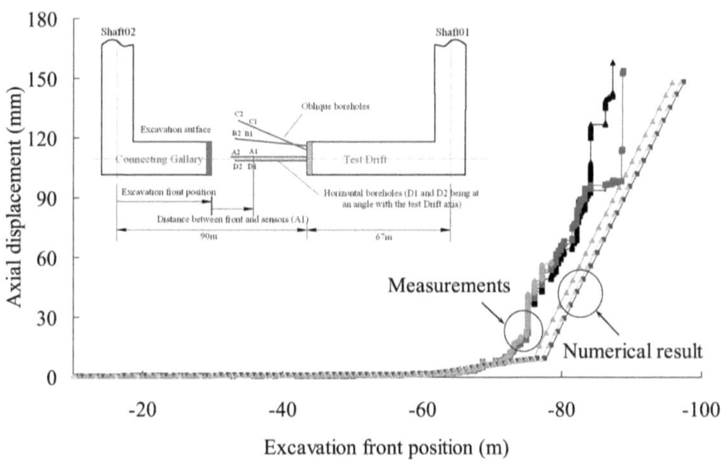

Fig. 6.14 Displacement comparisons between in situ measurements and predictions (A1)

6.6 Conclusions

This chapter presents a three-dimensional numerical simulation of the CLIPEX project conducted at the HADES underground research facility, aiming to elucidate the hydromechanical behavior of Boom Clay within the EDZ and HdZ. The simulated results demonstrate substantial agreement with in situ measurements, particularly in capturing the coupled hydromechanical response of the clay mass. Notably, the numerical modeling validates the theoretical frameworks developed in Chaps. 3 and 5, confirming their capacity to accurately predict both pressure distribution patterns and displacement evolution characteristics under complex underground conditions.

Chapter 7
3D Numerical Analysis of the In Situ PRACLAY THM Heating Test

7.1 Introduction

In situ heating experiments have been conducted at major underground research facilities worldwide to investigate the coupled thermo-hydro-mechanical (THM) responses of geological formations to heat sources simulating high-level radioactive waste disposal. These tests include the Drift Scale Test (DST) at Yucca Mountain, USA (Datta et al. 2004), the HE-D experiment at Mont Terri Underground Laboratory, Switzerland (Hoxha et al. 2006), the FEBEX installation at Grimsel URL, Switzerland (Alonso et al. 2005), the thermo-hydro-mechanical test at Kamaishi Mine, Japan (Chijimatsu et al. 2001; Rutqvist et al. 2001), the ATLAS (Chen et al. 2011, 2015; De Bruyn and Labat 2002; François et al. 2009), and PRACLAY Test at HADES URL, Belgium (Knowles et al. 1996; Picard et al. 1994; Van Marcke et al. 2013; Chen et al. 2021, 2024; Dizier et al. 2017, 2021). The experiments systematically measure temperature evolution, pore pressure dynamics, displacement patterns, and in situ stress variations under both backfilled and non-backfilled configurations (Gens et al. 2007; Bernier and Neerdael 1996).

Numerical simulation serves as a critical tool for interpreting the complex coupled THM processes in clay host formations during heating tests. Various numerical codes have been applied to model these experiments (Gens et al. 1998; Rutqvist and Tsang 2003; Zhang 2007; Charlier et al. 2010; Arson et al. 2010; Nowak et al. 2011; Chen et al. 2021). The PRACLAY in situ heating test was analyzed through three-dimensional numerical simulation to predict the long-term THM-coupled response of Boom Clay. The simulation incorporates the THMD model from Chap. 3 and the thermo-elasto-plastic creep damage model from Chap. 5, accounting for excavation disturbance effects, self-sealing mechanisms, and temperature-dependent hydraulic conductivity variations. The results provide predicted distributions of temperature fields, pore pressure gradients, creep strains, and damaged zone development across four operational phases: initial excavation, creep initiation, 10-year heating cycle, and subsequent 10-year cooling period.

W. Chen et al., *Thermo-Hydro-Mechanical Coupling in Clay Rock*,
https://doi.org/10.1007/978-981-95-2562-1_7

7.2 Brief Introduction of PRACLAY In Situ Heating Test

The PRACLAY in situ heating test ("PRACLAY Test") constitutes an active thermal research project implemented in the HADES URL (Fig. 6.1) at a depth of approximately 220 m within the Boom Clay formation (Bernier et al. 2007; Dizier et al. 2021). This experimental initiative aims to achieve three primary objectives: (1) quantify thermal impacts on thermo-hydro-mechanical (THM) coupled behavior, (2) evaluate clay sealing capacity under thermal loading, and (3) validate large-scale performance of Boom Clay under thermal stress conditions. The experimental configuration comprises three interrelated subsystems (Fig. 7.1): the gallery and crossing test section, the active heating test module, and the seal test. Operational since November 2014 (Chen et al. 2021). The PRACLAY in situ heating test is conducted within the PRACLAY gallery (PG), a cylindrical excavation measuring 45 m in length and 2.5 m external diameter, lined with precast concrete segments. The gallery features a 1.9 m internal diameter, resulting in 0.3 m thick structural linings. This primary test structure intersects orthogonally with the connecting gallery (CG). The active heating section extends along the gallery's central 30 m segment.

The PRACLAY gallery was excavated by an open-face tunneling machine at an excavation rate (excavation and installation of the lining) of 2 m/day in 2007. The gap between the initial radius of the gallery (over-excavation) and the extrados of the liner is about 0.06 m. The heater assembly, measuring 33 m in length, is positioned 10 cm radially inward from the liner's intrados (radius = 0.85 m). This system delivers a controlled, constant temperature of 80 °C at the gallery's extrados wall. A backfill material and hydraulic sealing system establish an "undrained" boundary condition. The annular bentonite layer applies a targeted swelling pressure 4.5 MPa against the Boom Clay formation. Figure 7.2 illustrates the hydraulic seal configuration.

The PRACLAY system incorporates an extensive instrumentation network comprising over 100 sensors (Fig. 7.3) specifically configured to acquire critical geotechnical parameters. These measurements include pore water pressure, relative

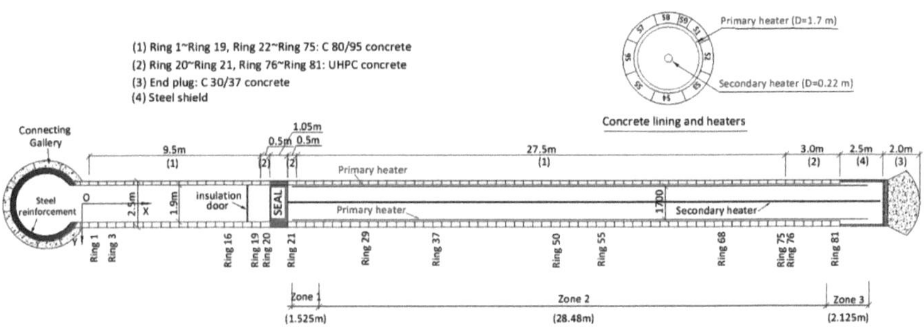

Fig. 7.1 Schematic representation of the PRACLAY in situ experimental setup (Chen et al. 2021)

Fig. 7.2 Schematic representation of the metallic seal assembly within the PRACLAY gallery (Dizier et al. 2021)

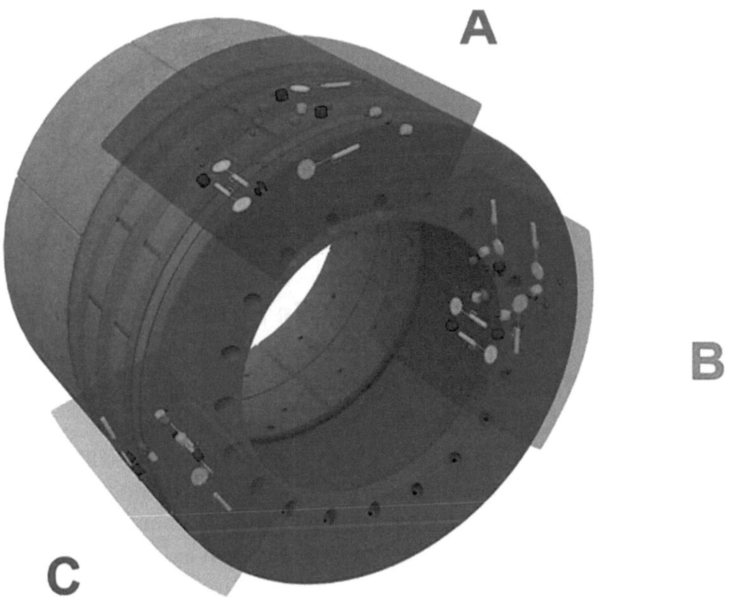

Fig. 7.3 Sensors grouped at three sections in the seal (yellow highlights indicate bentonite distribution within seal matrix) (Chen et al. 2021)

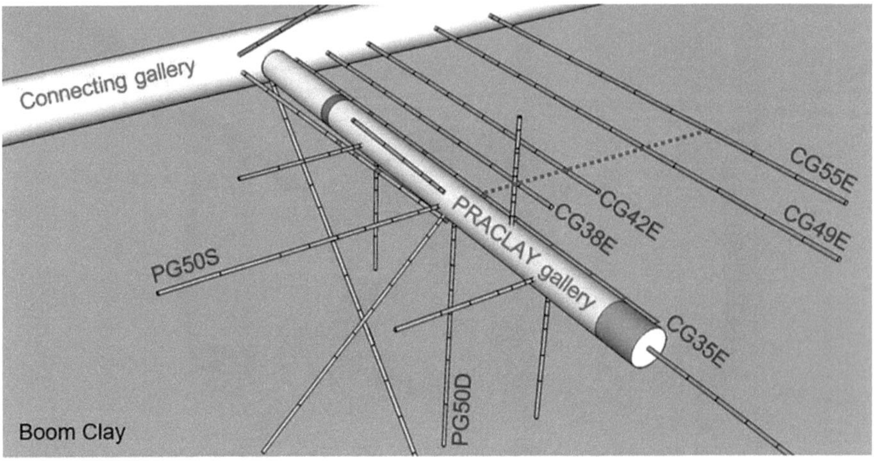

Fig. 7.4 Spatial arrangement of monitoring boreholes surrounding PRACLAY gallery (red dashed line is located in the mid-plane of the PRACLAY heater test) (Chen et al. 2021)

humidity, total stress, temperature, mechanical displacement, and axial strain, enabling real-time performance evaluation and constitutive behavior analysis.

The monitoring network consisted of three distinct subsystems: in situ clay massif instrumentation, reinforcement ring response monitoring, and structural health monitoring of the PRACLAY gallery lining. Five strategically positioned boreholes (CG35E, CG38E, CG42E, CG49E, and CG55E) were drilled parallel to the gallery's longitudinal axis at nominal radial distances of 2 m, 5 m, 9 m, 16 m, and 21 m, respectively. Figure 7.4 illustrates the spatial arrangement of monitoring boreholes surrounding the PRACLAY gallery.

7.3 3D FEM Model for PRACLAY In Situ Heating Test

7.3.1 Mesh Configuration

The 3D numerical model for the PRACLAY test was developed within a Cartesian coordinate system (O, x, y, z) as illustrated in Fig. 7.5, where the vertical z-axis follows upward orientation, while the x-axis aligns with the connecting gallery's longitudinal axis. The model origin coincides with the shaft axis at the gallery's depth of 223 m below the ground surface. Spatial discretization parameters include a longitudinal extent of 140 m (y-direction), lateral dimensions of 120 m (x and z directions), and a computational mesh comprising 398,061 C3D8RPT elements with 370,370 nodes. Symmetry conditions were applied across the vertical plane bisecting the gallery axis.

The numerical analysis incorporated realistic over-excavation effects through gap element modeling, accounting for the Boom Clay's pronounced time-dependent

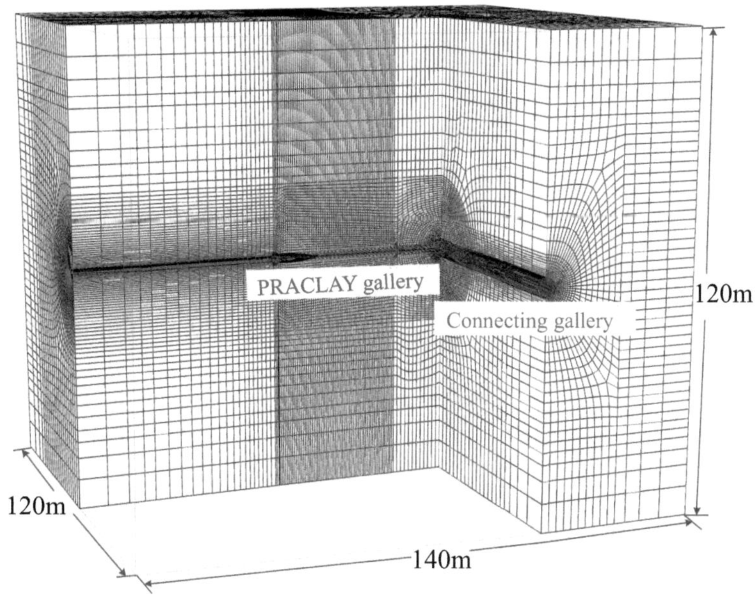

(a)

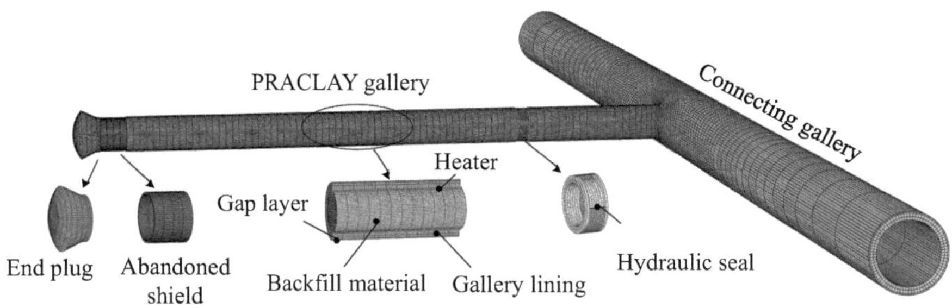

(b)

Fig. 7.5 3D FEM mesh in the modeling of PPRACLAY In situ Heating Test. (**a**) 3D view comprising underground structures and geological units; (**b**) gallery model

creep characteristics. As demonstrated in Fig. 7.6, this constitutive behavior induces progressive convergence in the concrete lining over operational timeframes, ultimately stabilizing under sustained loading conditions. The annular gap layer methodology (Fig. 7.7) enables simulation of shield-tunnel interaction through adjustable material properties, including elastic modulus modulation of gap elements and strain-controlled contact parameters. This dual-parameter control mechanism

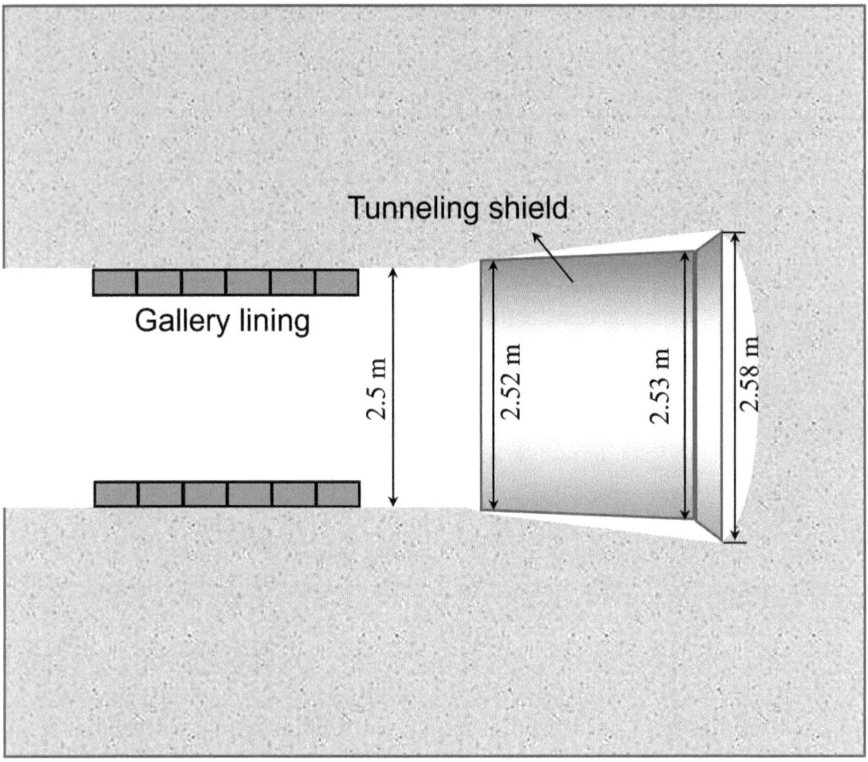

Fig. 7.6 Cross-section of the tunneling shield

effectively replicates interface behavior between the lining and surrounding geological formations under varying stress states.

7.3.2 *Initial and Boundary Conditions*

The numerical simulation was initialized with the following parameters:

7.3.2.1 Initial Conditions

- Vertical stress $\sigma_{zz} = \rho g z \approx 4.5$ MPa near gallery locations
- Horizontal stresses ($\sigma_{xx} = \lambda_x \sigma_{zz}$, $\sigma_{yy} = \lambda_y \sigma_{zz}$) in x and y directions
- Lateral total pressure coefficient ($\lambda_x = \lambda_y = 0.85$)
- Pore water pressure distribution following $u_w = \gamma_w z$
- Uniform initial temperature of 16.6 °C

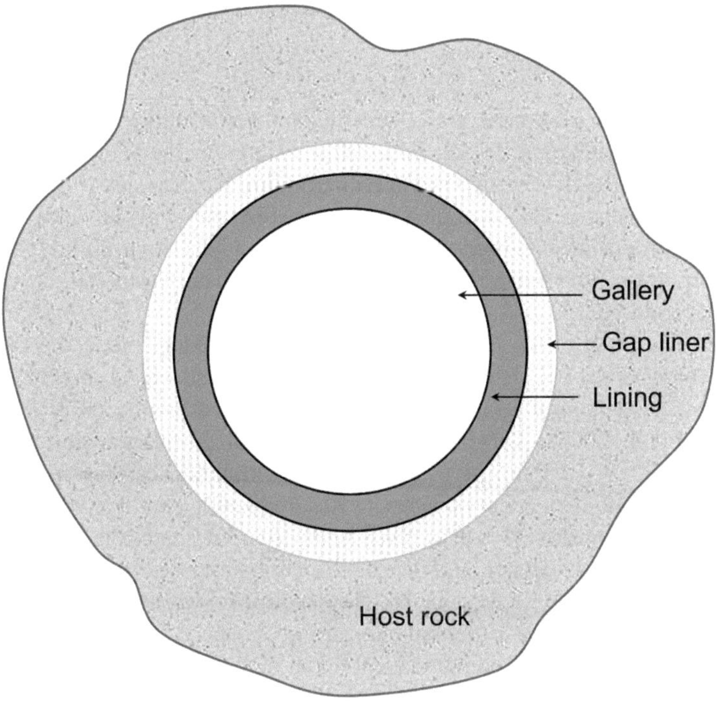

Fig. 7.7 The model of annular gap layer

7.3.2.2 Boundary Conditions

The model boundaries were configured as follows:

- Vertical stress application: Overburden pressure equivalent to stratigraphic weight applied at top surface.
- Deformation constraints: The normal deformation is fixed in the other boundaries.

Hydraulic Boundary:
- Impervious conditions maintained across all boundaries except lining intrados.
- Controlled pore pressure of 0.1 MPa at lining intrados during gallery opening phase.

Thermal Boundary:
- Convective heat transfer defined at lining intrados with $T\infty = 20$ °C and a thermal convection coefficient of about 5 W/(m^2 °C) on the lining intrados during the opening phase.
- Adiabatic conditions implemented on all other surfaces.

7.3.3 Hydraulic Conductivity Evolution Model for Boom Clay in EDZ/EdZ

The excavation process generates three distinct permeability alteration zones—excavation damaged zone (EDZ), excavation disturbed zone (EdZ), and hydraulic disturbed zone (HdZ)—through stress redistribution mechanisms (Fig. 7.8). While Sect. 3.5 established a THM-coupled model for HdZ, this study advances the methodology by proposing a comprehensive hydraulic conductivity evolution framework that incorporates both temperature and mechanical disturbance effects within the EDZ/EdZ continuum.

The construction of the HADES URL connecting gallery induced a two-order-of-magnitude increase in Boom Clay permeability (Bastiaens et al. 2003), as evidenced by the temporal–spatial evolution of hydraulic conductivity in Fig. 7.9 (Vervoort et al. 2005). This disturbance-driven enhancement manifested as an immediate $10^{-12} \rightarrow 10^{-10}$ m/s permeability surge within the near-field zone (<10 m), followed by a time-dependent recovery (~70% reduction over 3 years) attributed to self-sealing mechanisms. This dual-phase response necessitates the incorporation of both disturbance-induced permeability amplification and time-dependent self-sealing behavior into evolution models. Notably, high-level waste (HLW) disposal scenarios impose additional thermal constraints, where temperature-dependent permeability modulation becomes critical under repository operating conditions.

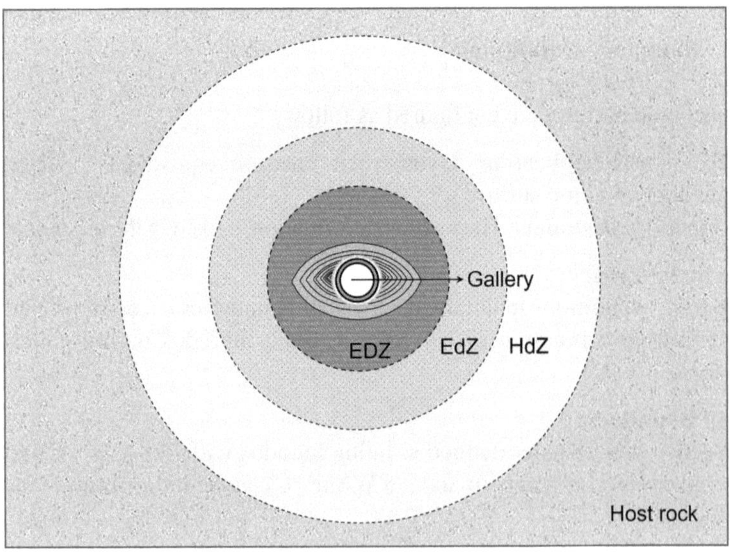

Fig. 7.8 Schematic diagram of EDZ, EdZ, and HdZ

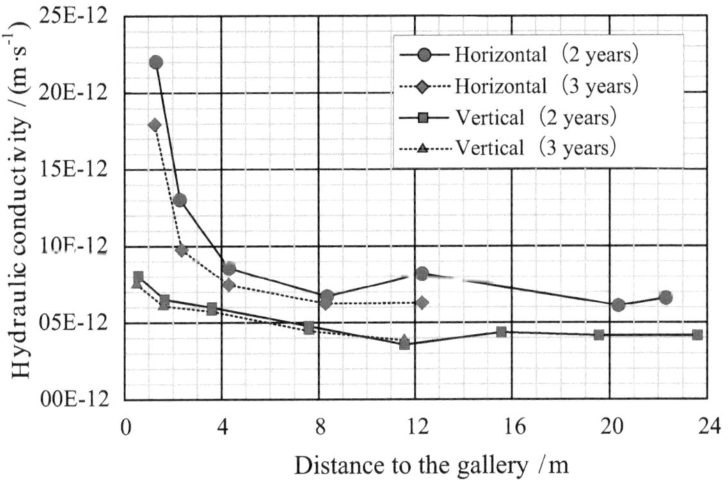

Fig. 7.9 Hydraulic conductivity around the connecting gallery at different times (Vervoort et al. 2005)

7.3.3.1 Hydraulic Conductivity Distribution Model for Boom Clay in the EDZ/EdZ

The progressive enhancement of intrinsic permeability within excavation-induced fracture zones (EDZ) and disturbance-affected zones (EdZ) originates from hierarchical fracture development driven by mineralogical heterogeneities, as substantiated by in situ pore pressure monitoring (Bastiaens et al. 2003). This study establishes the governing equation for hydraulic conductivity in these zones.

$$\begin{cases} K_{\mathrm{hd}} = K_{0\mathrm{h}} \cdot 10^{m \cdot y_{\mathrm{h}}} \\ y_{\mathrm{h}} = A_{\mathrm{h}} \cdot \exp\left(-r' / a_{\mathrm{h}}\right) + B_{\mathrm{h}} \end{cases} \tag{7.1}$$

$$\begin{cases} K_{\mathrm{vd}} = K_{0\mathrm{v}} \cdot 10^{m \cdot y_{\mathrm{v}}} \\ y_{\mathrm{v}} = A_{\mathrm{v}} \cdot \exp\left(-r' / a_{\mathrm{v}}\right) + B_{\mathrm{v}} \end{cases} \tag{7.2}$$

where K_{hd} and K_{vd} are the horizontal and vertical hydraulic conductivity of Boom Clay considering excavation disturbance in EDZ/EdZ, respectively; and $K_{0\mathrm{h}}$ and $K_{0\mathrm{v}}$ are the initial horizontal and vertical intrinsic permeability, respectively.

$$A_{\mathrm{h}} = \frac{1}{1 - \exp\left(-1 / a_{\mathrm{h}}\right)} \tag{7.3}$$

$$B_{\mathrm{h}} = \frac{-\exp\left(-1 / a_{\mathrm{h}}\right)}{1 - \exp\left(-1 / a_{\mathrm{h}}\right)} \tag{7.4}$$

$$A_{\mathrm{v}} = \frac{1}{1 - \exp(-1/a_{\mathrm{v}})} \tag{7.5}$$

$$B_{\mathrm{v}} = \frac{-\exp(-1/a_{\mathrm{v}})}{1 - \exp(-1/a_{\mathrm{v}})} \tag{7.6}$$

in which $r' = r/R_0$, r is the distance from center of the gallery, and R_0, $\underline{m}$, a_{h}, and a_{v} are material constants that need to be determined from field measurements of pore pressure.

7.3.3.2 Self-Sealing Model for Boom Clay

The self-sealing process of Boom Clay have been systematically investigated through surface laboratory and in situ monitoring (Van Marcke and Bastiaens 2010; Van Geet et al. 2008; Huysmans et al. 2006). This self-sealing phenomenon fundamentally arises from time-dependent interactions among clay mineral constituents, pore fluid chemistry, and mechanical stresses, constituting a tripartite coupling mechanism involving hydro-mechanical–chemical processes. The temporal evolution of fracture sealing is controlled by ion-exchange reactions, dissolution–precipitation kinetics, and stress-dependent microstructural reorganization under varying hydraulic gradients.

Therefore, this study proposes a hydro-geochemical sealing factor c_{seal}, mathematically formulated as:

$$c_{\mathrm{seal}} = \exp\left(-a_3 \frac{(p/f_0)^{\alpha'}}{(q'/f_0)^{\beta'}} t\right) \tag{7.7}$$

where a_3, α', β' are the material constants; p is the hydrostatic stress; q' is the Mises stress; $f_0 = 1$ MPa is the parameter for non-dimensionalization, and t is the sealing time.

Equation (7.7) quantitatively characterizes the hydro-mechanical coupling effects through two dimensionless parameters: α' representing the hydrostatic stress-enhanced self-sealing potential, and β' denoting the deviatoric stress-induced attenuation coefficient. This dual-parameter formulation establishes a mechanistic framework to decouple the competing hydrostatic strengthening and shear-induced weakening mechanisms.

The time-dependent evolution of hydraulic conductivity in fracture networks can thus be expressed as:

$$\begin{cases} K_{\mathrm{h}} = K_{0\mathrm{h}}\left[1 + \left(10^{\underline{m}\cdot y_{\mathrm{h}}} - 1\right)\cdot c_{\mathrm{seal}}\right] \\ K_{\mathrm{v}} = K_{0\mathrm{v}}\left[1 + \left(10^{\underline{m}\cdot y_{\mathrm{v}}} - 1\right)\cdot c_{\mathrm{seal}}\right] \end{cases} \tag{7.8}$$

where K_h and K_v are the horizontal and vertical hydraulic conductivity considering excavation disturbance and sealing effects in EDZ/EdZ, respectively.

7.3.3.3 Hydraulic Conductivity Evolution Model Considering Thermal Effect in EDZ/EdZ

The TIMODAZ European research consortium, focused on evaluating "Thermal Impact on the Damaged Zone around radioactive waste disposal in clay host rocks," has executed systematic experimental campaigns investigating thermal effects on clay self-sealing properties. Monfared et al. (2012) employed a custom-designed hollow cylindrical triaxial apparatus to analyze permeability evolution in sheared Boom Clay specimens under ambient (20 °C) and elevated (80 °C) temperature conditions. Chen et al. (2014) extended this research through temperature-variable permeability testing on fractured Boom Clay samples, complemented by micro-computed tomography (μCT) imaging to quantify thermal effects on fracture sealing dynamics. Despite these advancements, the precise thermo-mechanical coupling mechanisms governing temperature-dependent self-sealing in Boom Clay remain inconclusive. Consequently, building upon the THM-coupled permeability results from Chap. 2, this study incorporates thermally induced water viscosity and density variations into the hydraulic conductivity evolution model. The resultant formulation integrates three key mechanisms: (1) excavation-induced permeability enhancement (EDZ/EdZ), (2) time-dependent autogenous healing, and (3) thermo-mechanical coupling effects, expressed as:

$$\begin{cases} K_h(T) = K_{0h}(T_0)\left[1+\left(10^{\underline{m}\cdot y_h}-1\right)\cdot c_{seal}\right]\dfrac{\rho_w(T_0)\mu_w(T_0)}{\rho_w(T)\mu_w(T)} \\ K_v(T) = K_{0v}(T_0)\left[1+\left(10^{\underline{m}\cdot y_v}-1\right)\cdot c_{seal}\right]\dfrac{\rho_w(T_0)\mu_w(T_0)}{\rho_w(T)\mu_w(T)} \end{cases} \tag{7.9}$$

where $K_{0h}(T_0)$ and $K_{0v}(T_0)$ are the initial horizontal and vertical intrinsic permeability at ambient temperature, respectively.

The calibrated governing equation parameters for hydraulic conductivity evolution models are presented in Table 7.1, derived through coupled back analysis of in situ pore pressure monitoring data from the two excavation galleries.

Table 7.1 Parameters for hydraulic conductivity due to excavation and self-sealing

Items	R_0	$\underline{m}$	β'	α'	a_h	a_v	a_3
Values	30	2.121	0.1012	0.5852	0.334	0.795	0.1669

7.3.4 THM Coupling Parameters for the Model

Comprehensive parameter documentation for the hydraulic conductivity evolution model (Table 7.1), THMD model (Table 3.3), and thermo-elasto-plastic creep damage model (Tables 5.1 and 5.2) are systematically presented, with key thermo-hydro-mechanical properties of Boom Clay summarized in Chap. 3. As established in Chap. 5, the CREEP subroutine in ABAQUS inherently restricts anisotropic elastic parameter implementation, necessitating the adoption of isotropic elastic modulus in this numerical simulation. The comprehensive THM parameter framework for underground structures is additionally detailed in Table 7.2.

7.3.5 Modeling Procedure

The numerical simulation framework, grounded in saturated porous media theory, systematically simulates the THM response of the HADES URL construction sequence through 10 sequential phases (Fig. 7.10).

7.3.5.1 Phase 1: Connecting Gallery Excavation (45 d)

Full-section excavation simulation implemented via section failure method, with concurrent activation of gap elements and lining elements under drained boundary conditions. Pore pressure maintained at 1.0 bar on lining intrados to prevent excessive pore pressure accumulation. Initial geothermal gradient preserved (15 °C) throughout this phase.

Table 7.2 THM parameters for underground structures

Parameter	Concrete	Steel	Backfill sand
Young's modulus/GPa	44	300	1.0
Poison ratio	0.25	0.25	0.25
Dry density/kg m^{-3}	2100	6000	2100
Void ratio	0.176	–	0.66
Biot coefficient	1.0	–	1.0
Intrinsic permeability/m^2	4×10^{-18}	–	10^{-8}
Specific heat capacity/J kg^{-1} K^{-1}	250	1400	880
Thermal conductivity/W m^{-1} K^{-1}	2.06	10	2.1
Thermal expansion coefficient/K^{-1}	10^{-5}	1.3×10^{-5}	10^{-5}

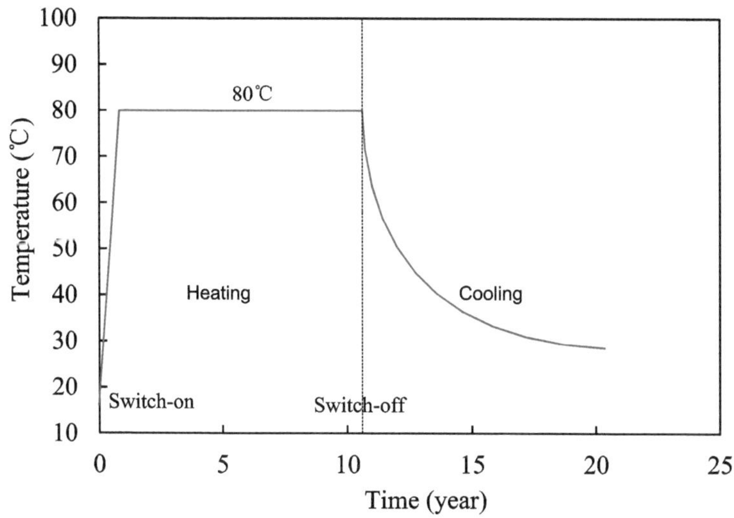

Fig. 7.10 Temperature evolution of the heater

7.3.5.2 Phase 2: Post-Excavation Drainage (5.5 a)

Sustained pore pressure boundary (1.0 bar) applied to connecting gallery lining intrados to simulate natural consolidation processes.

7.3.5.3 Phase 3: PRACLAY Gallery Excavation (33 d)

Identical full-section excavation methodology applied as Phase 1, maintaining structural continuity.

7.3.5.4 Phase 4: Dual-Gallery Drainage (2.2 a)

Combined drainage boundary conditions applied to both galleries, with additional thermal convection boundary (ambient temperature 20 °C, heat transfer coefficient 5 W/(m^2 °C)) imposed on PRACLAY lining intrados.

7.3.5.5 Phase 5: Hydraulic Seal Installation (70 d)

Removal of designated lining segments and substitution with bentonite-steel composite barrier system.

7.3.5.6 Phase 6: Interim Backfill Stabilization (1.7 a)

Maintenance of dual-gallery drainage conditions during interim period prior to backfill emplacement.

7.3.5.7 Phase 7: Backfill Material Pressurization (168 d)

Progressive pressure application to sand backfill material, incrementing from 1.0 bar to 5.0 bar through staged loading protocol.

7.3.5.8 Phase 8: Heater Pre-Pressurization (1.65 a)

Controlled pore pressure increase from 5 bar to 8 bar within backfill material to simulate operational stress conditions.

7.3.5.9 Phase 9: Thermal Loading Cycle (10.5 a)

Biphasic thermal regime: Initial 6-month linear temperature ramp (heater surface temperature from initial state to 80 °C). Subsequent 10-year maintenance phase at 80 °C interface temperature between Boom Clay and lining.

7.3.5.10 Phase 10: Natural Cooling (10 a)

Removal of active thermal boundary conditions, permitting conductive heat dissipation from heater assembly through Boom Clay massif.

The temperature evolution of the heater during the heating and cooling phase is shown in Fig. 7.10.

7.4 Analysis of the Numerical Results

The locations of cross-sections for view cut are shown in Fig. 7.11.

7.4.1 Temperature and Pore Pressure Profile Before the Heating Stage

The pore water pressure distribution around the Connecting gallery before PRACLAY gallery excavation is presented in Fig. 7.12. The analysis reveals significant alterations in the surrounding rock pore pressure due to excavation activities

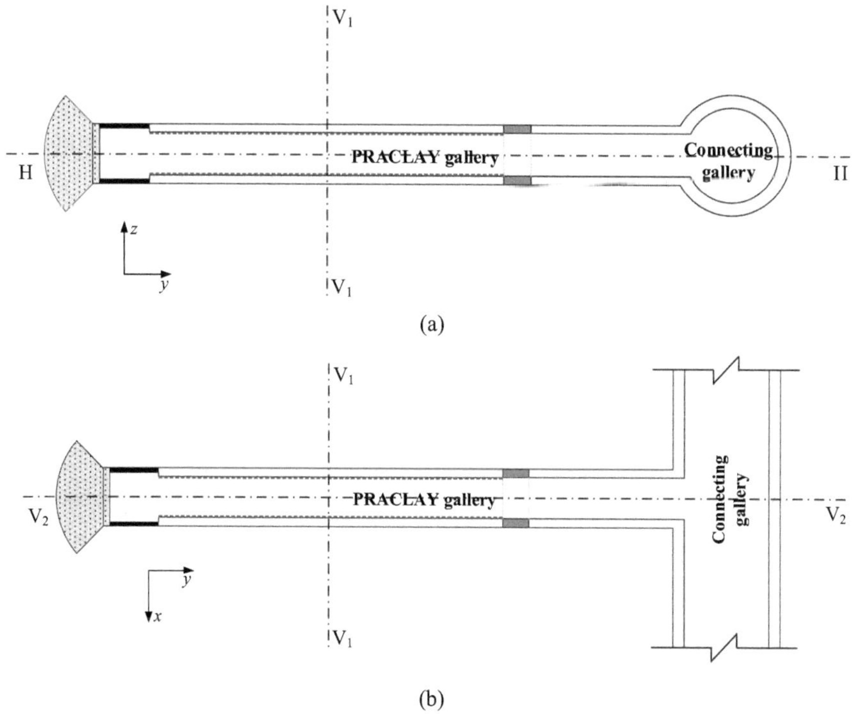

Fig. 7.11 Locations of cross-sections for view cut

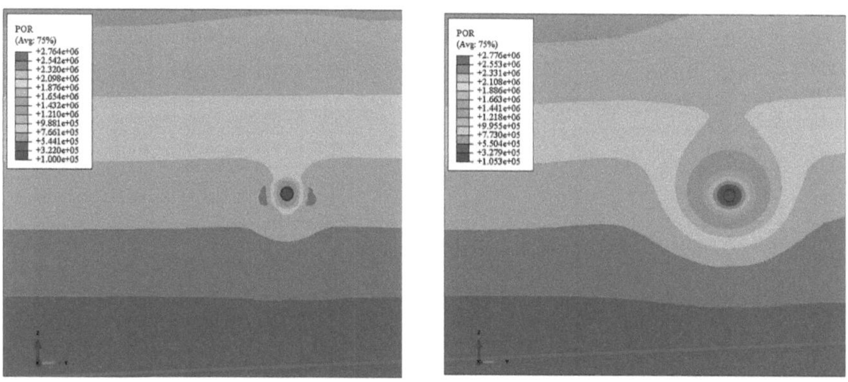

(a) After Connecting gallery excavation

(b) 5.5 years after excavation

Fig. 7.12 Pore water pressure profile around connecting gallery before the excavation of PRACLAY gallery on V_2–V_2 plane (unit: Pa). (**a**) After connecting gallery excavation; (**b**) 5.5 years after excavation and subsequent creep deformation. Horizontal pressure attenuation exceeded vertical dissipation, demonstrating pronounced anisotropic response.

As illustrated in Fig. 7.13, the temporal evolution of pore pressure around the PRACLAY gallery exhibits analogous characteristics. A pore pressure anomaly of 2.5–2.6 MPa developed at 2–3 m horizontally from the gallery wall, exceeding the initial pore pressure. This phenomenon, attributed to the formation's anisotropic mechanical properties, was successfully reproduced by the numerical model. Vertical pressure dissipation patterns were observed concurrent with horizontal pressure accumulation. Notably, backfill material installation and pressurization significantly influenced pressure development, as evidenced by post-pressurization pressure increments near the gallery walls (Figs. 7.13c, d). Before thermal testing,

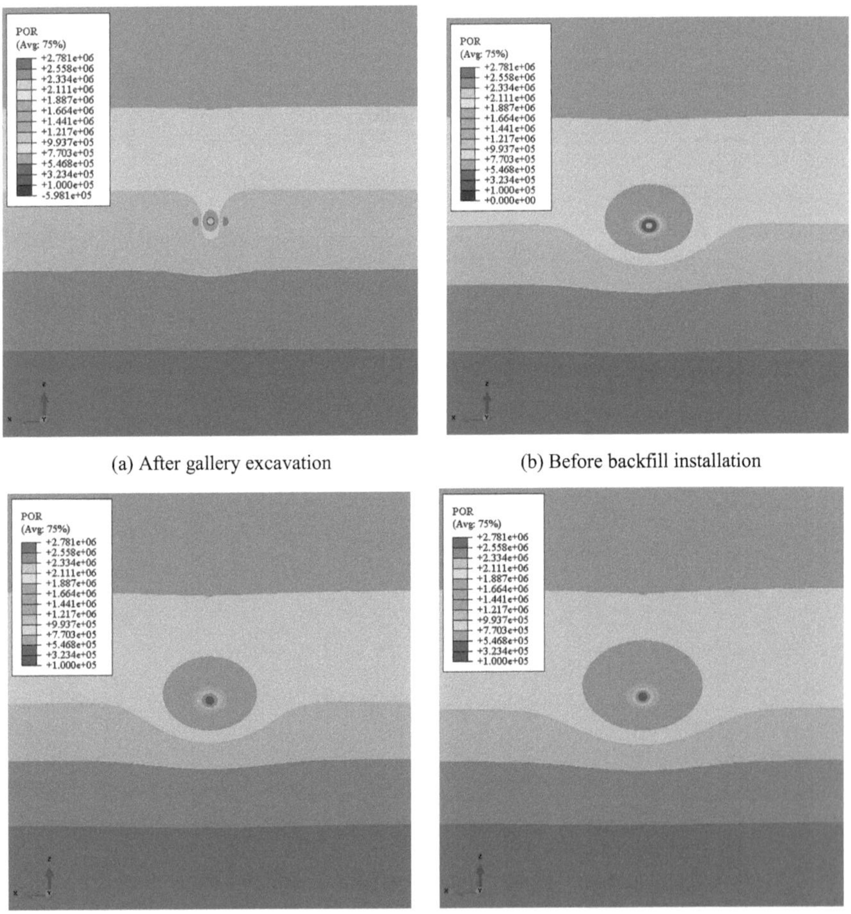

(a) After gallery excavation (b) Before backfill installation

(c) After pressurizing the backfilled sand to 0.5 MPa (d) Before the heating test

Fig. 7.13 Distribution of pore pressure around PRACLAY gallery on V1–V1 plane (unit: Pa). (**a**) After gallery excavation; (**b**) before backfill installation; (**c**) after pressurizing the backfilled sand to 0.5 MPa; (**d**) before the heating test

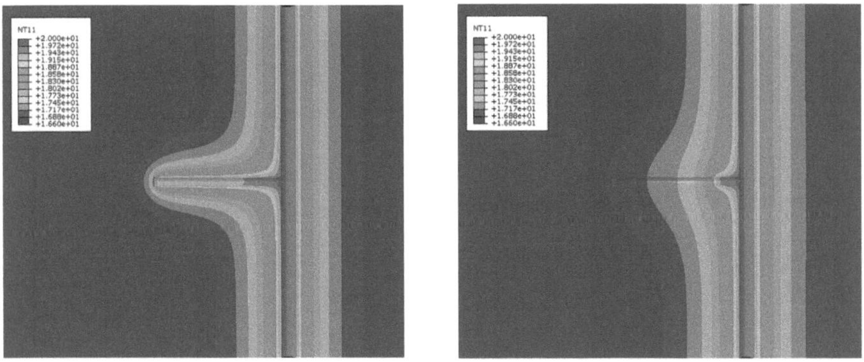

(a) Before backfilling (b) Before heating

Fig. 7.14 Temperature variation in Boom Clay due to ventilation before the heating test. (**a**) Before backfilling; (**b**) before heating

intrados pore pressures had risen to approximately 9 bar, indicating substantial hydraulic response to backfill consolidation.

A convective heat transfer boundary condition was imposed at the intrados of the gallery lining system. Progressive thermal accumulation within the Boom Clay formation occurred over operational timeframes due to sustained ventilation, resulting in maximum temperature elevation at the structural intersection between PRACLAY and connecting galleries. Thermal mapping revealed distinct spatial gradients, with the lateral thermal influence zone extending approximately 30 m horizontally from the connecting gallery axis, while vertical propagation reached 25 m. Pre-heating baseline thermal conditions are documented in Fig. 7.14, demonstrating the baseline geothermal gradient prior to thermal stimulation.

7.4.2 *Evolution of Temperature Distribution During Heating–Cooling Stages*

Figures 7.15 and 7.16 document the thermal evolution in the surrounding Boom Clay during the heating stimulation phase, presented as vertical (Fig. 7.15) and horizontal (Fig. 7.16) temperature profiles. The thermal influence footprint extended approximately 35–40 m radially from the heating chamber following a 10.5-year heating period. Comparative analysis revealed distinct spatial dissimilarities between vertical and horizontal temperature gradients, with contour patterns exhibiting subtle elliptical characteristics attributed to differential thermal diffusivity anisotropy, specifically enhanced horizontal thermal conductivity relative to vertical orientation.

Figure 7.17 presents the thermal response in horizontal (H) and vertical (V) profiles across the V1–V1 plane during the heating phase. Pronounced thermal gradients were observed during the initial heating stage, with rapid temperature

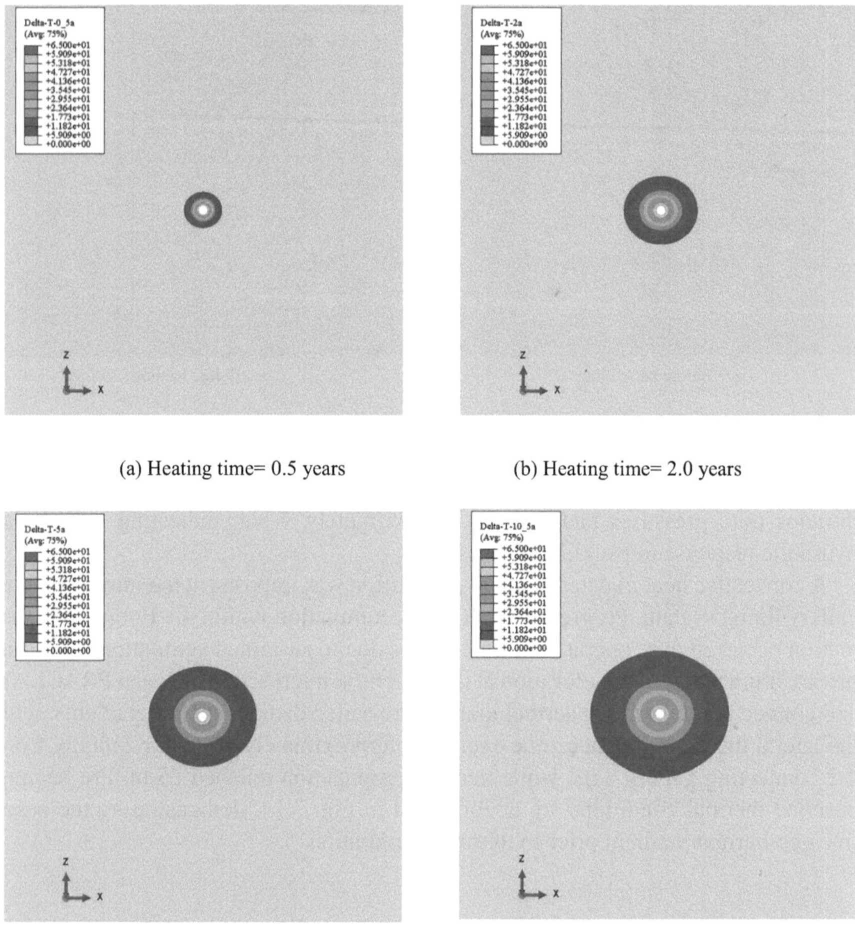

(a) Heating time= 0.5 years (b) Heating time= 2.0 years

(c) Heating time= 5.0 years (d) Heating time= 10.5 years

Fig. 7.15 Temperature evolution during the heating stage on V_1–V_1 plane (unit: °C). (**a**) Heating time = 0.5 years; (**b**) heating time = 2.0 years; (**c**) heating time = 5.0 years; (**d**) heating time = 10.5 years

propagation near the gallery lining during the first year. This transient period transitioned to stabilized thermal regimes from year 5 through year 10, demonstrating diminished temperature diffusion rates over prolonged operational durations. The anisotropic thermal response aligned with previous observations of horizontal thermal conductivity exceeding vertical values, as evidenced by elliptical isotherm patterns in planar projections.

Figures 7.18 and 7.19 document the temperature evolution in the surrounding Boom Clay during the 10-year cooling phase, analyzed through vertical (Fig. 7.18) and horizontal (Fig. 7.19) profiles. Immediate thermal contraction occurred near the

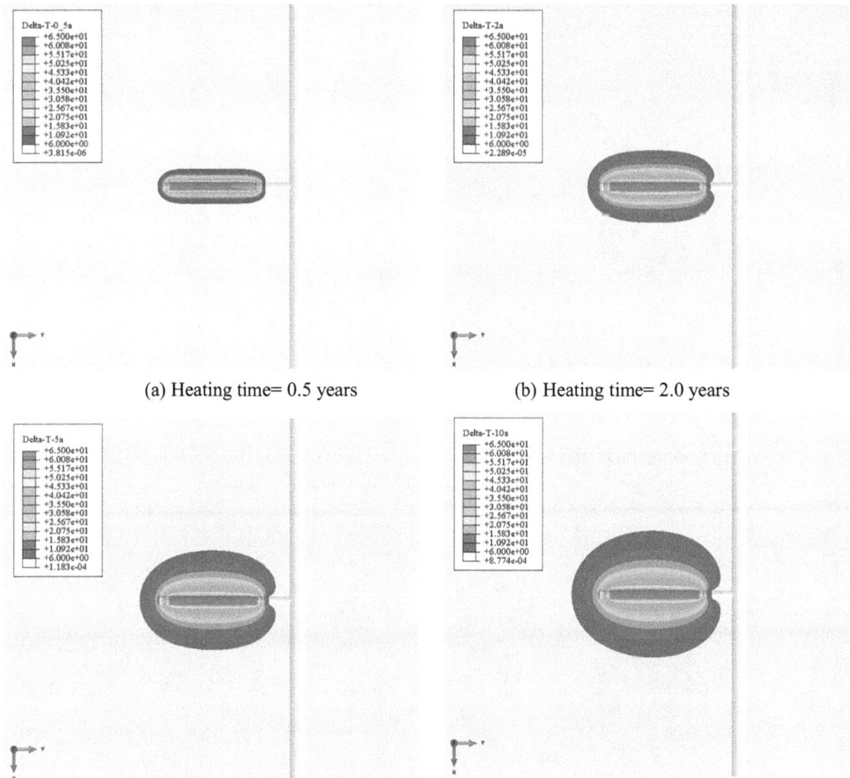

(a) Heating time= 0.5 years (b) Heating time= 2.0 years

(c) Heating time= 5.0 years (d) Heating time= 10.5 years

Fig. 7.16 Temperature evolution during the heating stage in H–H plane (unit: °C). (**a**) Heating time = 0.5 years; (**b**) heating time = 2.0 years; (**c**) heating time = 5.0 years; (**d**) heating time = 10.5 years

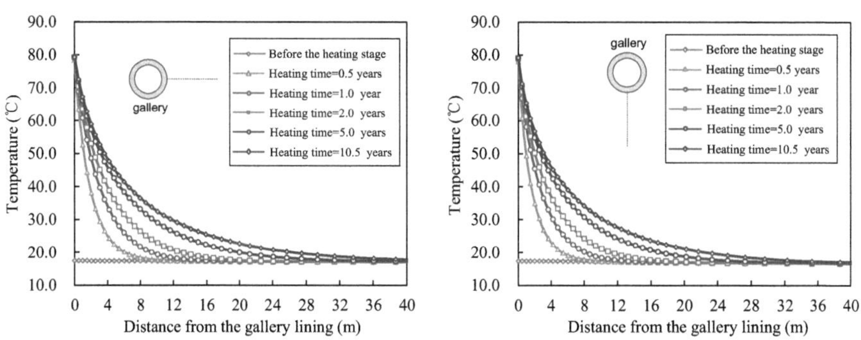

(a) Horizontal profile (b) Vertical profile

Fig. 7.17 Temperature evolution in different directions around PRACLAY gallery (V_1–V_1 plane). (**a**) Horizontal profile; (**b**) vertical profile

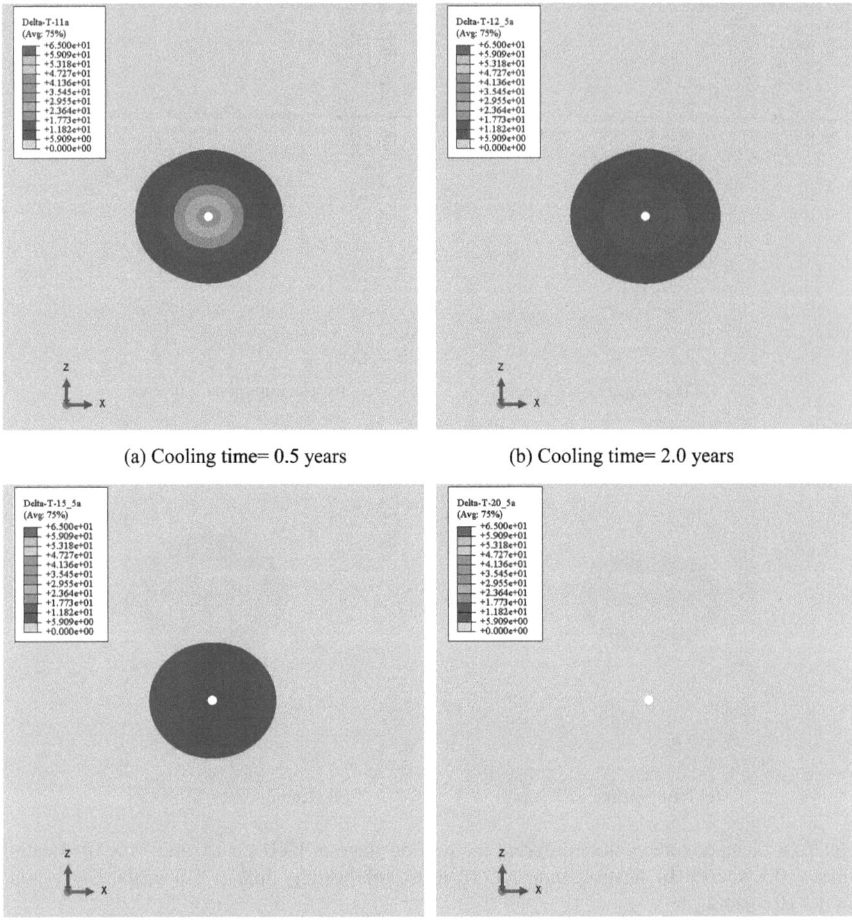

(a) Cooling time= 0.5 years (b) Cooling time= 2.0 years

(c) Cooling time = 5.0 years (d) Cooling time = 10.0 years

Fig. 7.18 Temperature evolution during the cooling stage on V_1–V_1 plane (unit: °C). (**a**) Cooling time = 0.5 years; (**b**) cooling time = 2.0 years; (**c**) cooling time = 5.0 years; (**d**) cooling time = 10.0 years

gallery lining post-heater shutdown, whereas distal zones exhibited transient thermal inertia—initial temperature elevation followed by gradual decline. Comparative analysis of planar responses in Fig. 7.20 (V_1–V_1 plane) confirms this dichotomous behavior, with both horizontal and vertical temperature trajectories converging to <22 °C across near-field (<5 m) and far-field (>30 m) domains after full cooling stabilization.

Figure 7.21 presents the thermal response progression recorded by instrumented boreholes across the median plane. The simulated temperature regimes will undergo rigorous calibration against empirical measurements acquired during subsequent

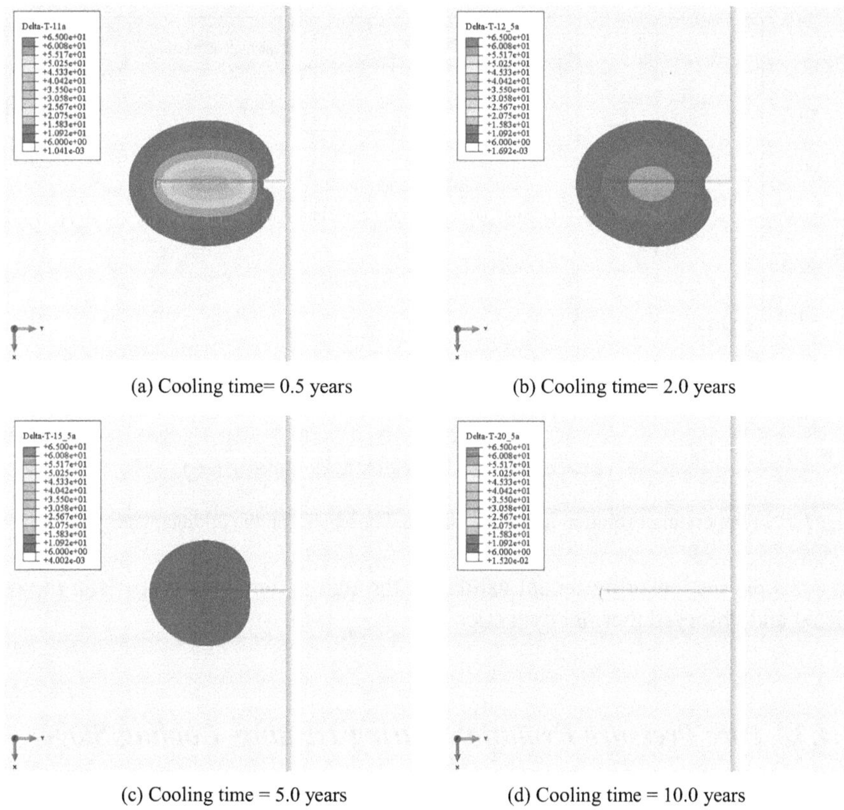

Fig. 7.19 Temperature evolution during the cooling stage on H–H plane (unit: °C). (**a**) Cooling time = 0.5 years; (**b**) cooling time = 2.0 years; (**c**) cooling time = 5.0 years; (**d**) cooling time = 10.0 years

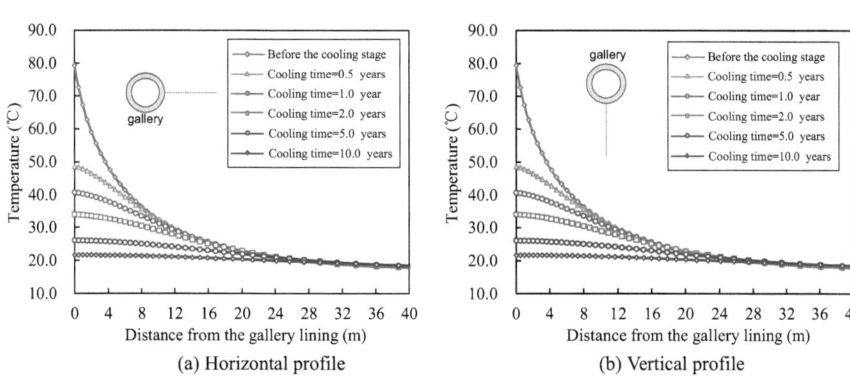

Fig. 7.20 Temperature evolution in different directions around PRACLAY gallery (V_1–V_1 plane). (**a**) Horizontal profile; (**b**) vertical profile

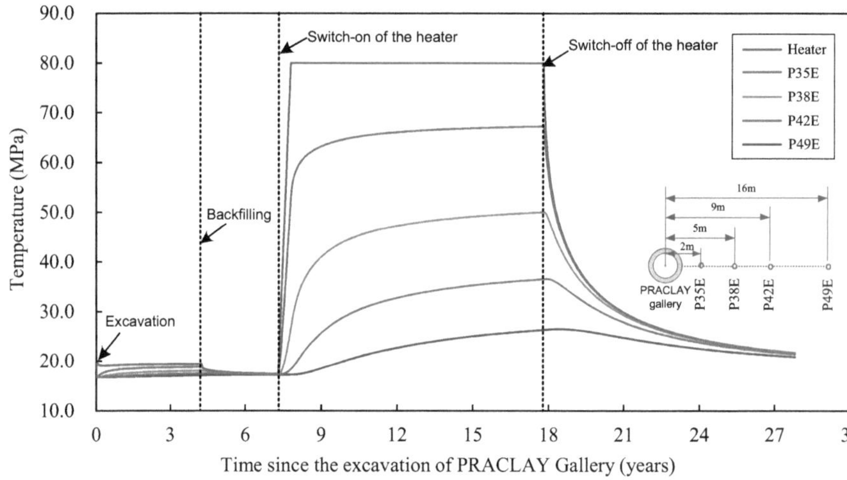

Fig. 7.21 Temperature evolution in the instrumented boreholes on the middle plane

analysis phases, ensuring model validation through systematic comparison of predicted and observed thermal dynamics.

7.4.3 Pore Pressure Evolution During Heating–Cooling Stages

Figures 7.22 and 7.23 present spatiotemporal distributions of pore pressure variation in vertical (Fig. 7.22) and horizontal (Fig. 7.23) planes during heating phase operations. The observed hydraulic response emerges from competing mechanisms: (1) thermal pressurization driven by differential thermal expansion of fluid–solid matrix and (2) pressure dissipation through aquifer systems. Initial heating phases induced rapid pore pressure accumulation (up to 3.25 MPa) within the heated zone, constrained by undrained boundary conditions imposed by backfill sand and hydraulic containment systems. Elliptical pressure isocontours (major axis $\approx$1.2$\times$ minor axis) reflect pronounced thermal conductivity anisotropy in Boom Clay formations.

Figure 7.24 documents the spatiotemporal evolution of pore water pressure distribution in horizontal and vertical profiles across the V_1–V_1 plane. The analysis reveals progressive pressure dissipation from the clay-lining interface to distal zones, with peak pressure migration correlating with thermal contraction mechanisms.

Figure 7.25 presents the seepage velocity distribution around the hydraulic seal (heating duration = 0.5 year). The results demonstrate elevated seepage velocities ($\approx$1.7$\times$ background rates) near the seal interface, establishing hydraulic gradient amplification as a critical factor for pressure dissipation adjacent to gallery linings.

Figures 7.26 and 7.27 document the post-cooling pore pressure evolution in response to thermal contraction mechanisms. The analysis reveals rapid pressure

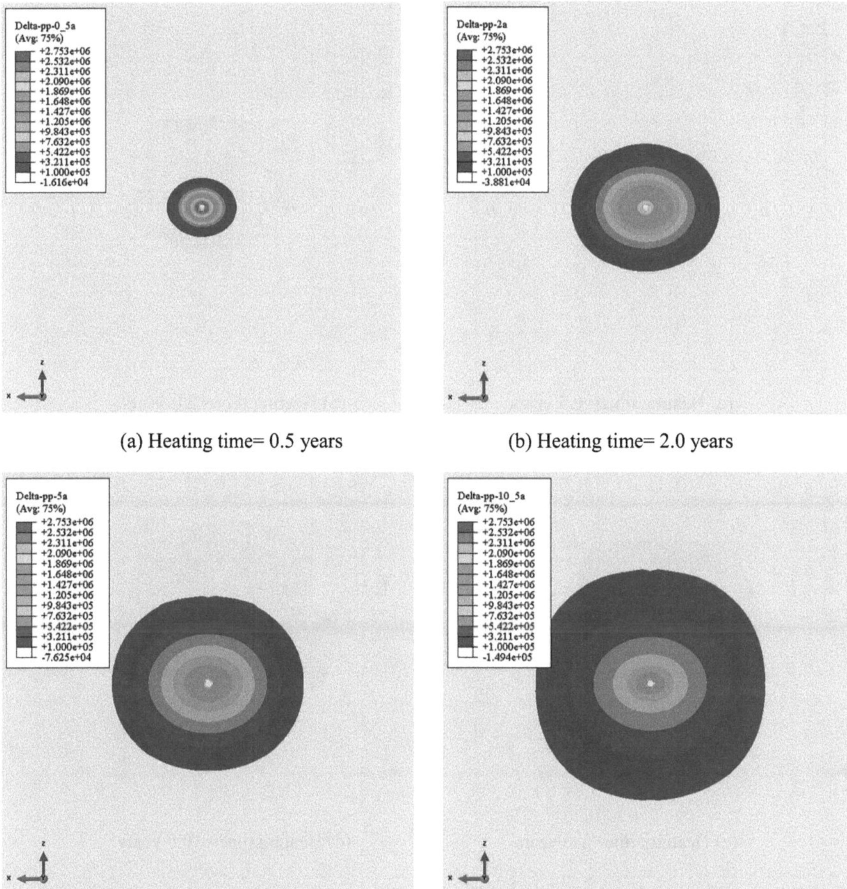

(a) Heating time= 0.5 years

(b) Heating time= 2.0 years

(c) Heating time= 5.0 years

(d) Heating time= 10.5 years

Fig. 7.22 Pore pressure evolution during the heating stage on V_1–V_1 plane (unit: Pa). (**a**) Heating time = 0.5 years; (**b**) heating time = 2.0 years; (**c**) heating time = 5.0 years; (**d**) heating time = 10.5 years

dissipation during initial cooling phases, with near-field pressure reduction exceeding background levels to establish new hydraulic equilibrium states. Progressive thermal recovery demonstrates progressive pressure restoration toward pre-heating baseline conditions through sustained thermal equilibration processes.

Figure 7.28 shows the pore pressure profile in the horizontal and vertical directions of the V_1–V_1 plane.

Figure 7.29 illustrates the temporal evolution of pore pressure distribution observed in instrumented boreholes located on the median plane. The analysis demonstrates delayed pressure response characteristics in far-field zones compared to near-field regions, with sustained pressure accumulation persisting even after 10 years of postheating recovery.

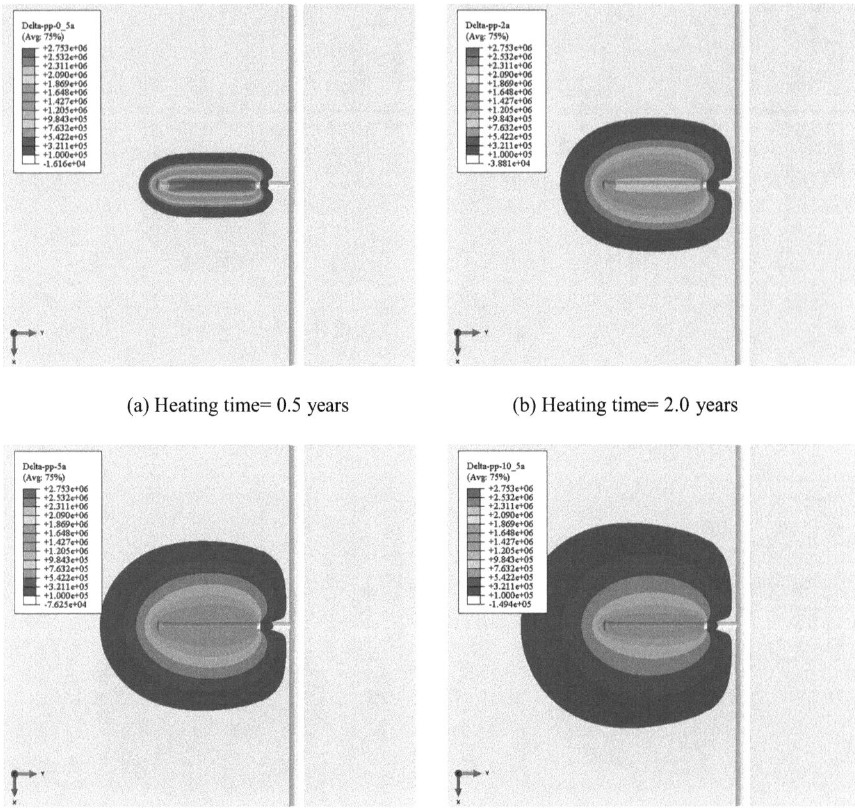

(a) Heating time= 0.5 years (b) Heating time= 2.0 years

(c) Heating time= 5.0 years (d) Heating time= 10.5 years

Fig. 7.23 Pore pressure evolution during the heating stage in H–H plane (unit: Pa). (**a**) Heating time = 0.5 years; (**b**) heating time = 2.0 years; (**c**) heating time = 5.0 years; (**d**) heating time = 10.5 years

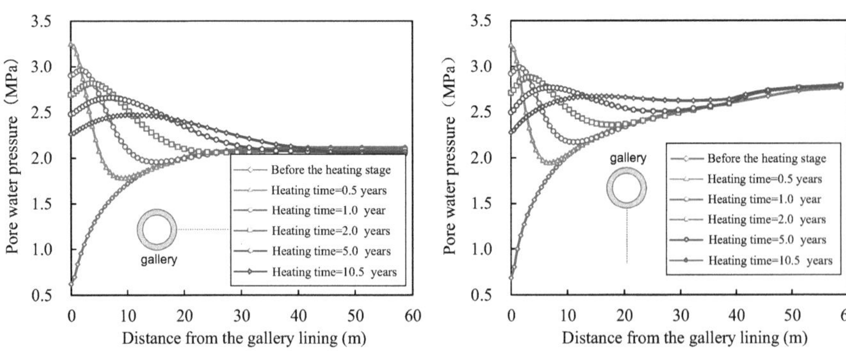

(a). Horizontal profile (b).Vertical profile

Fig. 7.24 Pore water pressure distribution around PRACLAY gallery at different times (V_1–V_1 plane). (**a**) Horizontal profile; (**b**) vertical profile

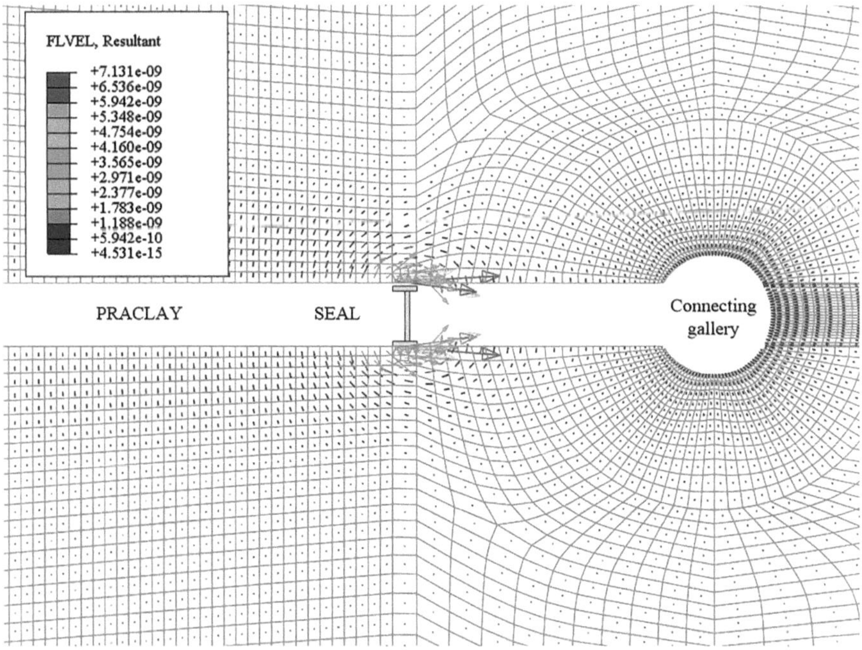

Fig. 7.25 The seepage velocity distribution around the hydraulic seal (heating time = 0.5 years)

7.4.4 Creep Strain Development During Heating–Cooling Stage

Figure 7.30 presents the spatial distribution of equivalent creep strain variations through successive thermal phases in the horizontal plane. Pre-heating baseline conditions revealed accumulated creep strains of 12.5 years for the surrounding Boom Clay and 7.0 years for PRACLAY gallery clay. Maximum strain concentration (6.84×10^{-3}) occurred at the gallery intersection. Enhanced creep activity was documented near heating chambers during active thermal stimulation, while cooling phases exhibited minimal strain evolution due to reduced deformation rates.

Figure 7.31 documents the temporal distribution of equivalent creep strain variations in the vertical plane across successive thermal phases. Baseline conditions revealed pre-existing creep strains of 1.5×10^{-3} in horizontal PRACLAY gallery strata before thermal stimulation. Enhanced creep activity was observed in Boom Clay adjacent to the heating chamber during thermal stimulation, with maximum strain concentration reaching 4.7×10^{-3} prior to cooling initiation. Subsequent cooling phases exhibited reduced deformation rates while maintaining residual strain accumulation, ultimately stabilizing at 5.2×10^{-3} at experimental termination.

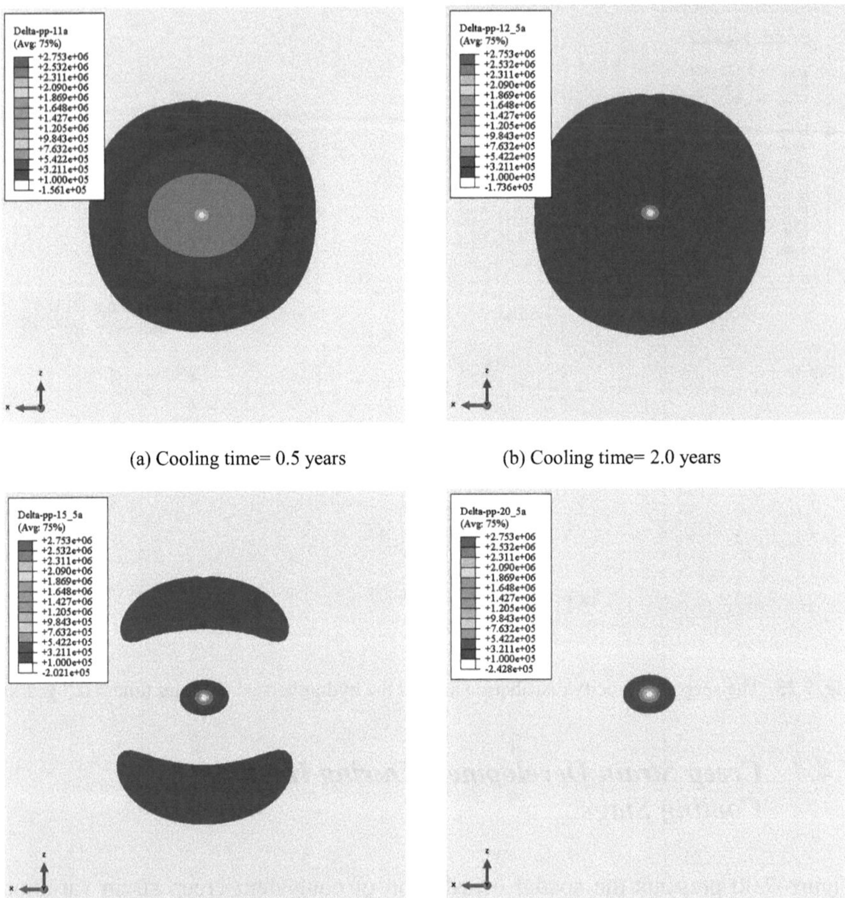

(a) Cooling time= 0.5 years (b) Cooling time= 2.0 years

(c) Cooling time = 5.0 years (d) Cooling time = 10.0 years

Fig. 7.26 Pore pressure evolution during the cooling stage on V_1–V_1 plane (unit: Pa). (**a**) Cooling time = 0.5 years; (**b**) cooling time = 2.0 years; (**c**) cooling time = 5.0 years; (**d**) cooling time = 10.0 years

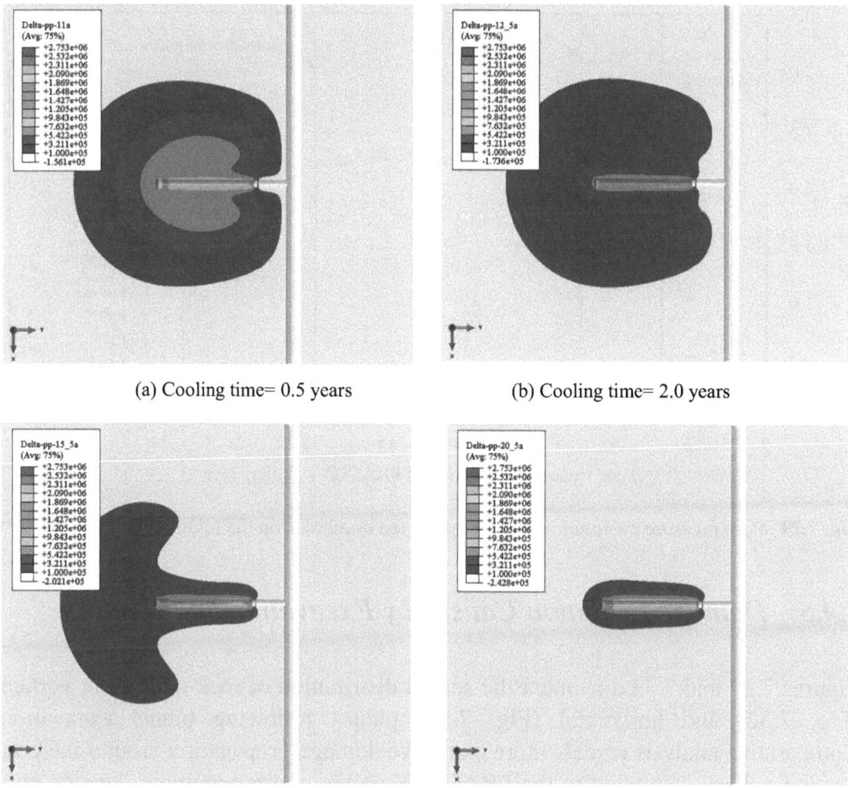

(a) Cooling time= 0.5 years (b) Cooling time= 2.0 years

(c) Cooling time = 5.0 years (d) Cooling time = 10.0 years

Fig. 7.27 Pore pressure evolution during the cooling stage in H–H plane (unit: Pa). (**a**) Cooling time = 0.5 years; (**b**) cooling time = 2.0 years; (**c**) cooling time = 5.0 years; (**d**) cooling time = 10.0 years

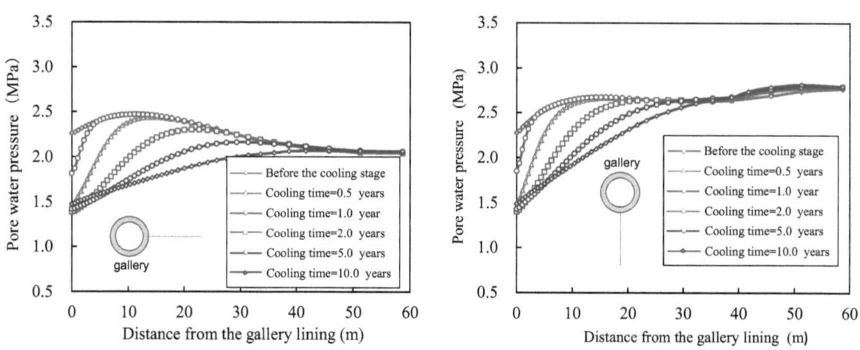

(a) Horizontal profile (b) Vertical profile

Fig. 7.28 Pore water pressure distribution around PRACLAY gallery at different times (V_1–V_1 plane). (**a**) Horizontal profile; (**b**) vertical profile

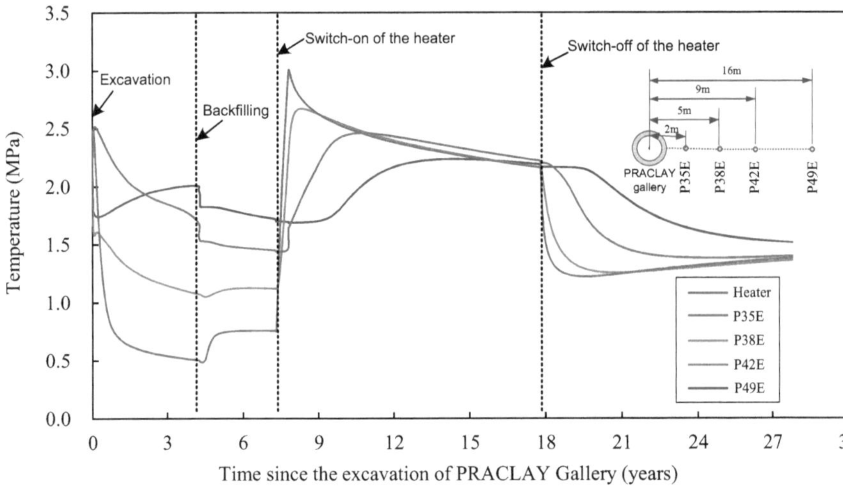

Fig. 7.29 Pore pressure evolution in the instrumented boreholes on the middle plane

7.4.5 *Damage Evolution Caused by Excavation and Heating*

Figures 7.32 and 7.33 document the spatial distribution of rock damage in vertical (Fig. 7.32) and horizontal (Fig. 7.33) planes following tunnel excavation. Comparative analysis reveals more extensive damage propagation around the connecting gallery compared to the PRACLAY gallery, with maximum damage concentration occurring at the intersection zone due to stress concentration effects. Pre-heating creep conditioning resulted in progressive damage expansion within surrounding strata, exhibiting greater horizontal damage extent attributed to anisotropic creep behavior.

Figures 7.34 and 7.35 present time-lapse damage evolution through successive thermal phases. Heating stimulation induced thermal damage accumulation, causing damage zone expansion. Notably, post-heating cooling phases demonstrated limited damage progression despite continued creep deformation, indicating thermal damage saturation.

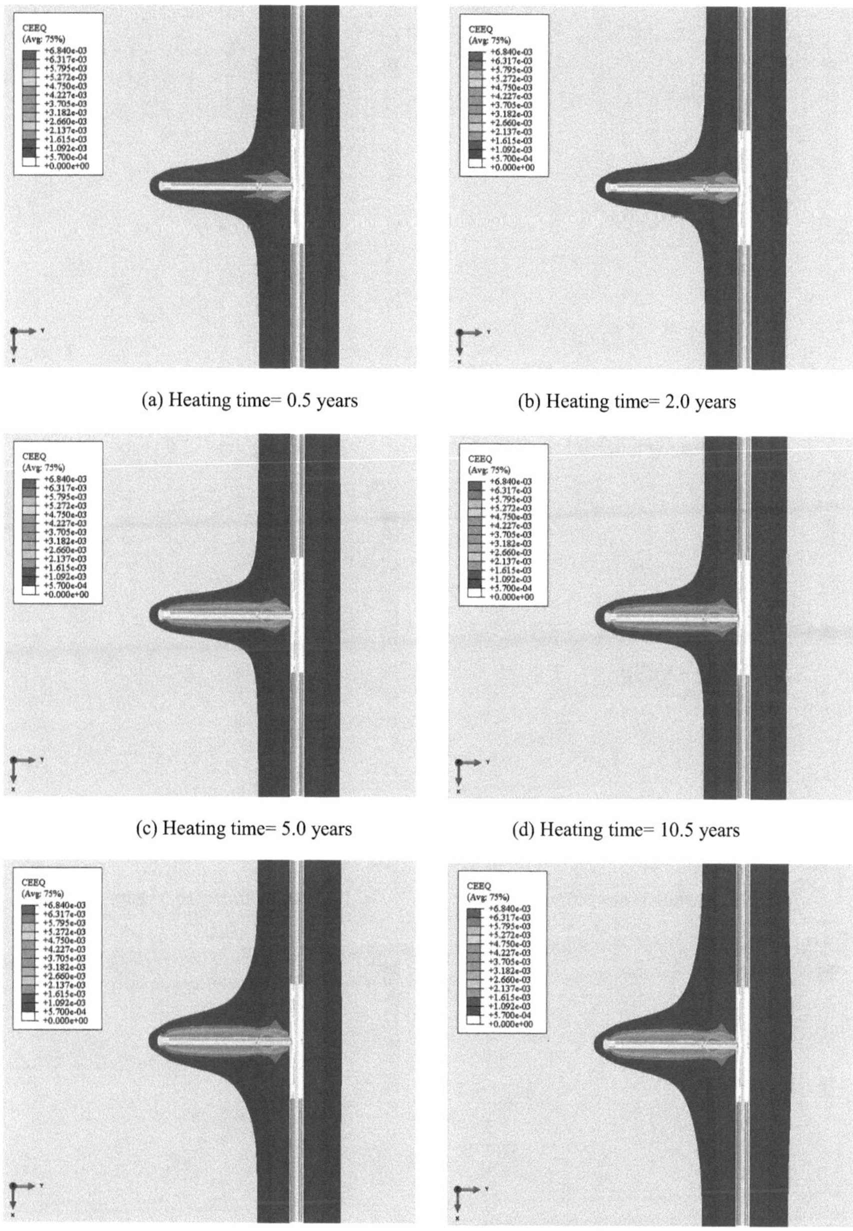

(a) Heating time= 0.5 years (b) Heating time= 2.0 years

(c) Heating time= 5.0 years (d) Heating time= 10.5 years

(e) Cooling time = 5.0 years (f) Cooling time = 10.0 years

Fig. 7.30 The evolution of equivalent creep strain in H–H plane. (**a**) Heating time = 0.5 years; (**b**) heating time = 2.0 years; (**c**) heating time = 5.0 years; (**d**) heating time = 10.5 years; (**e**) cooling time = 5.0 years; (**f**) cooling time = 10.0 years

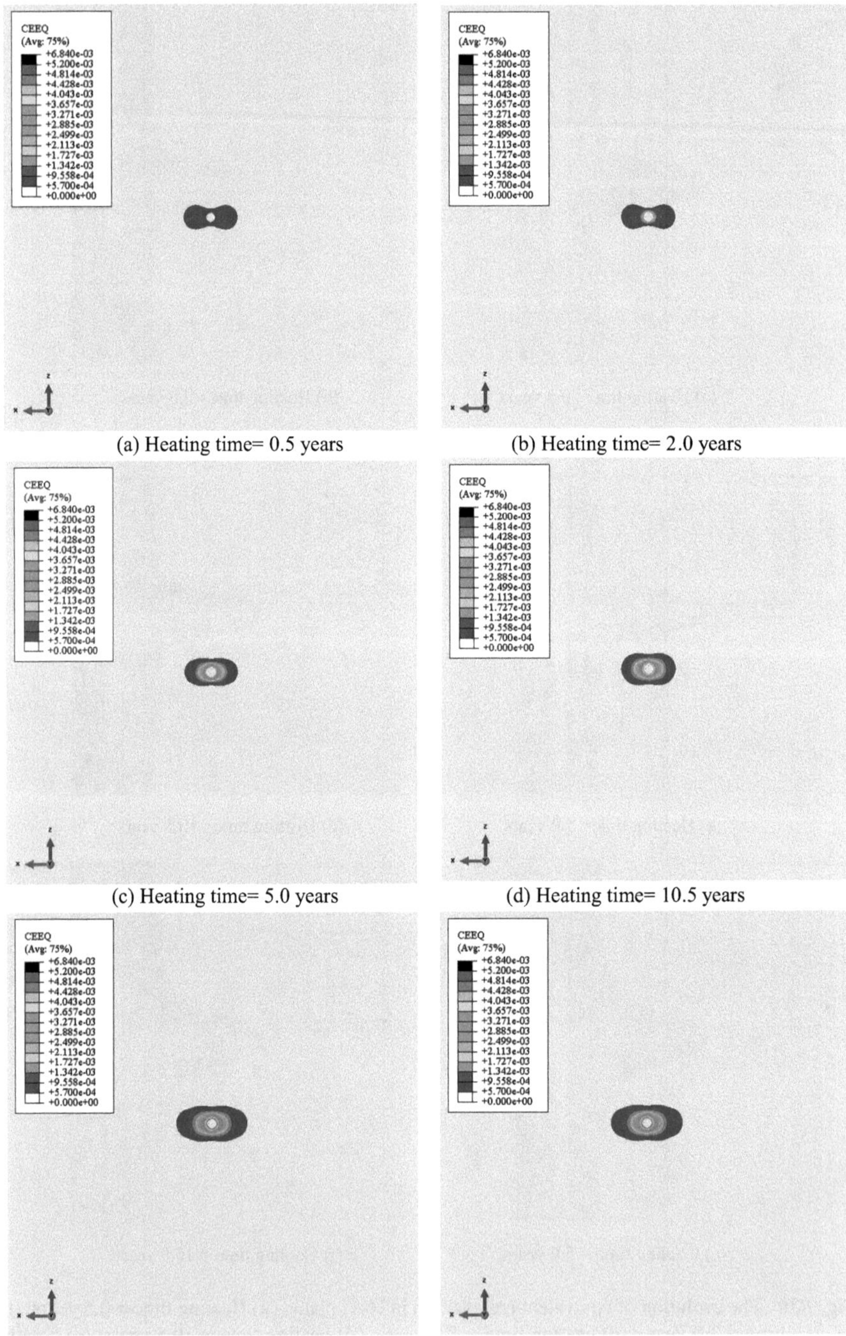

Fig. 7.31 The evolution of equivalent creep strain in V_1-V_1 plane. (**a**) Heating time = 0.5 years; (**b**) heating time = 2.0 years; (**c**) heating time = 5.0 years; (**d**) heating time = 10.5 years; (**e**) cooling time = 5.0 years; (**f**) Cooling time = 10.0 years

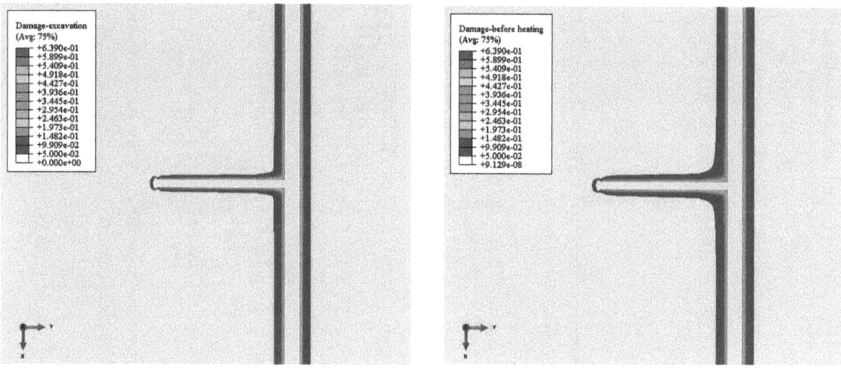

(a) After excavation of PRACLAY (b) Before heating

Fig. 7.32 Profile of damage around the galleries before heating in H–H plane. (**a**) After excavation of PRACLAY; (**b**) before heating

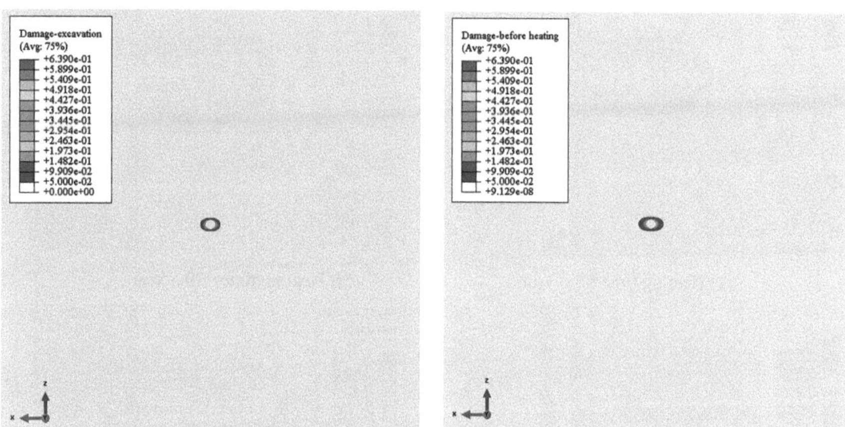

(a) After excavation of PRACLAY (b) Before heating

Fig. 7.33 Profile of damage around the galleries before heating in V_1–V_1 plane. (**a**) After excavation of PRACLAY; (**b**) before heating

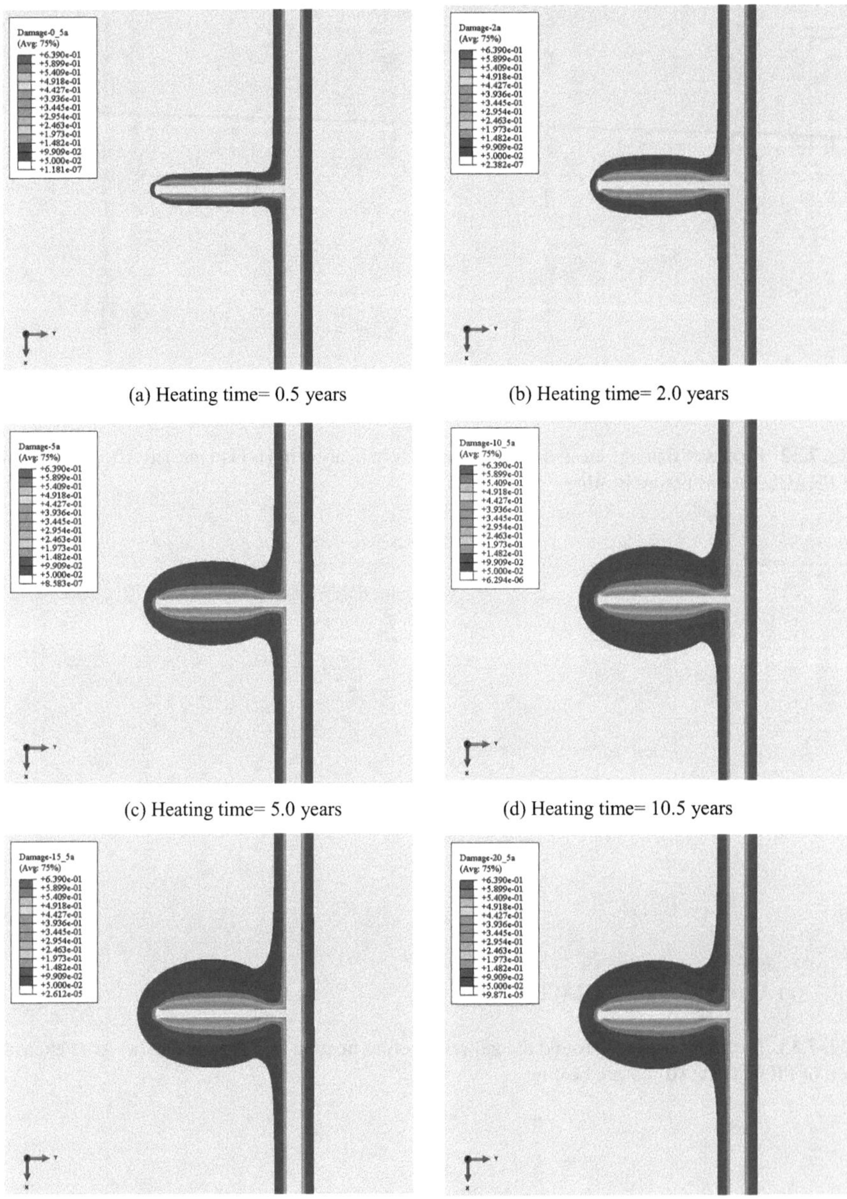

(a) Heating time= 0.5 years (b) Heating time= 2.0 years

(c) Heating time= 5.0 years (d) Heating time= 10.5 years

(e) Cooling time = 5.0 years (f) Cooling time = 10.0 years

Fig. 7.34 The evolution of damage in H–H plane. (**a**) Heating time = 0.5 years; (**b**) heating time = 2.0 years; (**c**) heating time = 5.0 years; (**d**) heating time = 10.5 years; (**e**) cooling time = 5.0 years; (**f**) cooling time = 10.0 years

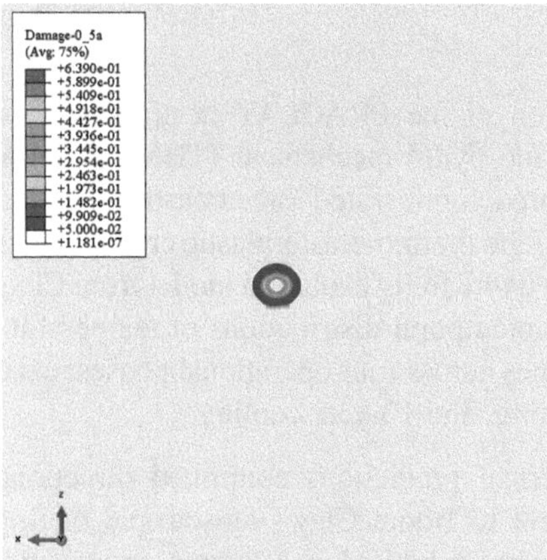

(a) Heating time= 0.5 years

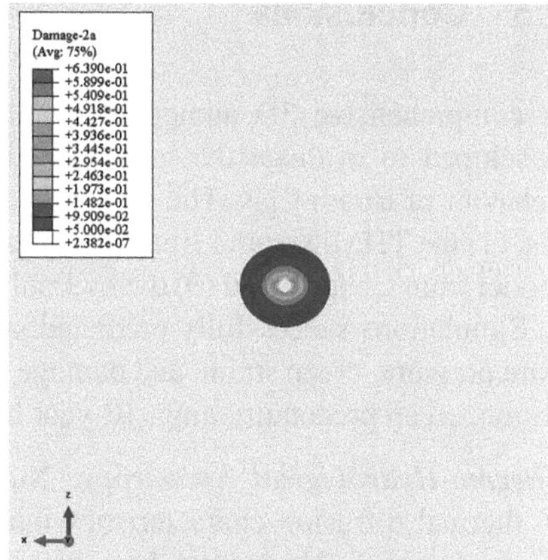

(b) Heating time= 2.0 years

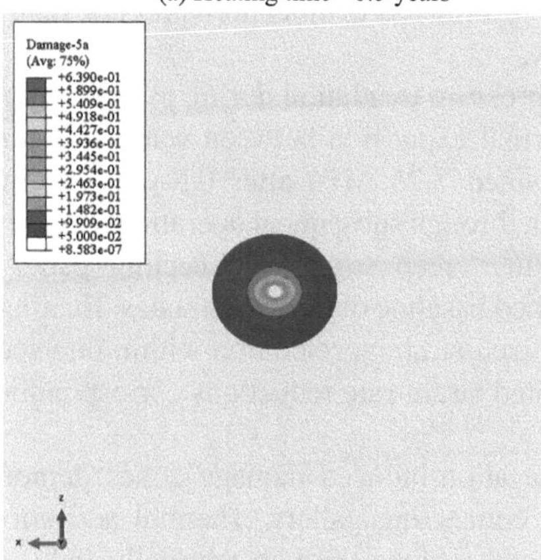

(c) Heating time= 5.0 years

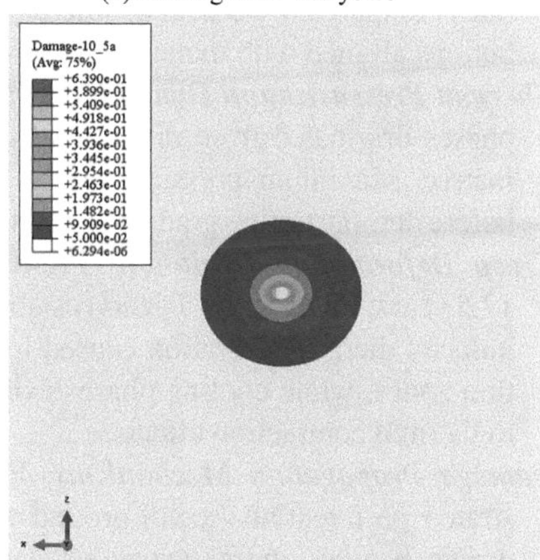

(d) Heating time= 10.5 years

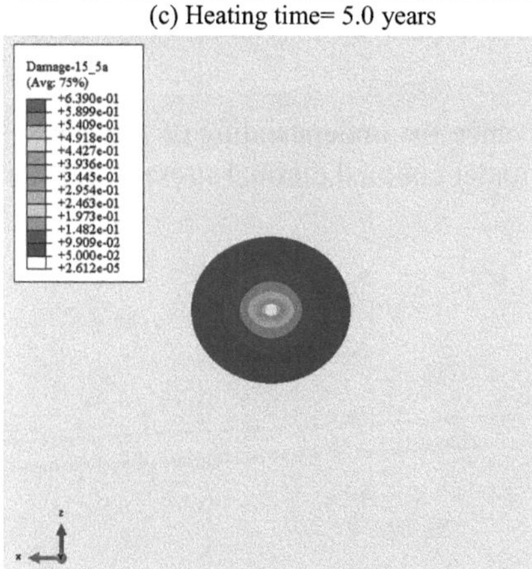

(e) Cooling time = 5.0 years

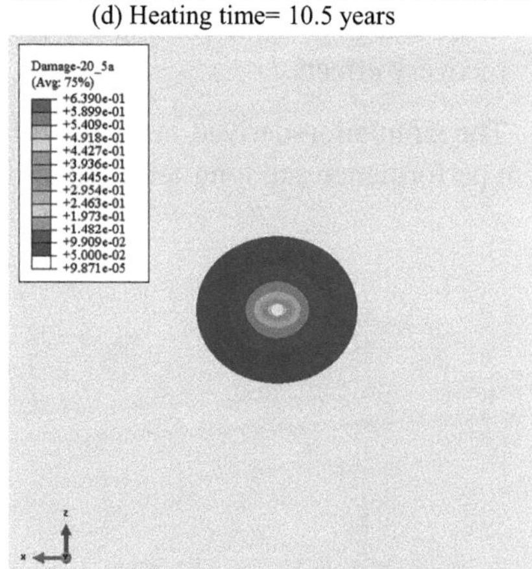

(f) Cooling time = 10.0 years

Fig. 7.35 The evolution of damage in V_1–V_1 plane. (**a**) Heating time = 0.5 years; (**b**) heating time = 2.0 years; (**c**) heating time = 5.0 years; (**d**) heating time = 10.5 years; (**e**) cooling time = 5.0 years; (**f**) cooling time = 10.0 years

7.5 Conclusions

A comprehensive 3D numerical simulation of the PRACLAY heating test was developed to evaluate the long-term thermo-hydro-mechanical (THM) coupling behavior of Boom Clay. The modeling approach integrated three constitutive models: (1) the THMD model from Chap. 3, (2) the thermo-elasto-plastic creep damage model from Chap. 5, and (3) the hydraulic conductivity evolution model from Chap. 7. Simulations successfully predicted spatiotemporal distributions of temperature, pore pressure, creep strain, and damage zones across four operational phases: excavation, creep preconditioning, 10-year heating, and 10-year cooling.

Thermo-Hydrological Anisotropy: Numerical predictions confirmed directional thermal diffusion characteristics inherent to Boom Clay's anisotropic thermal conductivity. Horizontal heat propagation exhibited accelerated propagation rates compared to vertical counterparts, resulting in distinct pore pressure distributions aligned with material anisotropy.

Thermal Pressurization Dynamics: Pore pressure escalation during initial heating phases originated from differential thermal expansion between water and clay matrix. Maximum pressure values attained 3.25 MPa after 0.5-year heating before demonstrating gradual attenuation through subsequent operational phases.

Creep Deformation Evolution: Pre-existing creep strains (connecting gallery: 12.5 years; PRACLAY: 7 years) established baseline deformation states. Heating-induced thermal activation caused localized strain acceleration within intersection zones, while cooling phases exhibited strain-rate reductions corresponding to thermal contraction effects.

Damage Propagation Mechanisms: Excavation-induced damage zones demonstrated preferential growth around the connecting gallery. Thermal activation during heating phases expanded damaged regions through thermally induced microcracking, while cooling contributions remained constrained by viscoelastic recovery effects.

The simulation-derived insights will advance the understanding of sensing system performance and long-term reliability under coupled thermal stress conditions.

References

Abaqus FEA. ABAQUS analysis user's manual[M]. Dassault systemes. France: Vélizy-Villacoublay; 2006.

Abuel-Naga HM, Bergado DT, Ramana GV, et al. Experimental evaluation of engineering behavior of soft Bangkok clay under elevated temperature. J Geotech Geoenviron. 2006;132(7):902–10.

Abuel-Naga HM, Bergado DT, Bouazza A, et al. Volume change behavior of saturated clays under drained heating conditions: experimental results and constitutive modeling. Can Geotech J. 2007;44(8):942–56.

Abuel-Naga HM, Bergado DT, Bouazza A, et al. Thermomechanical model for saturated clays. Géotechnique. 2009;59(3):273–8.

Al Reda SM, Yu C, Berthe G, Matray JM. Study of the permeability in the Opalinus clay series (Mont Terri-Switzerland) using the steady state method in Hassler cell. J Pet Sci Eng. 2020;184:106457.

Almanza R, Castañeda R, Silva G. Temperature-electrolyte effects on clay soil liners[C]. In: Proceedings of the 1st international conference on unsaturated soils, vol. 1; 1995. p. 343–8.

Alonso EE, Alcoverro J, Coste F, et al. The FEBEX benchmark test: case definition and comparison of modelling approaches. Int J Rock Mech Min Sci. 2005;42(5):611–38.

Armand G, Leveau F, Nussbaum C, de La Vaissiere R, Noiret A, Jaeggi D, et al. Geometry and properties of the excavation-induced fractures at the Meuse/haute-Marne URL drifts. Rock Mech Rock Eng. 2014;47(SI):21–41.

Arson C, Charlier R, Chambon R, et al. TIMODAZ report: deliverable D10—constitutive models development, TIMODAZ; 2010.

Arulanandan K, Shen CK, Young RB. Discussion: undrained creep behaviour of a coastal organic silty clay. Géotechnique. 1972;22(3):536–7.

Baldi G, Borsetto M and Hueckel T. Calibration of mathematical models for the simulation of thermal, seepage and mechanical behaviour of boom clay. CEC Publishers, Luxembourg. EUR 10924 EN. 1987, 215.

Baldi G, Hueckel T, Pellegrini R. Thermal volume changes of the mineral-water system in low-porosity clay soils. Can Geotech J. 1988;25(4):807–25.

Baldi G, Hueckel T, Peano A, Pellegrini R. Further check and development of the thermo-hydro-geomechanical model for boom clay and clay based buffer material with an emphasis on thermo-mechanics of water solid system. Bergamo: ISMES, WAS; 1990. Final report; 380.83.7.I(S)

Baldi G, Hueckel T, Peano A, and Pellegrini R. Developments in modelling of thermo-hydro-mechanical behaviour of boom clay and clay based buffer materials. CEC Publishers, Luxembourg, EUR. 13365 EN, 1991, vol. 3, p. 134.

W. Chen et al., *Thermo-Hydro-Mechanical Coupling in Clay Rock*,
https://doi.org/10.1007/978-981-95-2562-1

Bastiaens W. EDZ around an industrial excavation in Boom Clay. In: Davies C, Bernier F, editors. EDZ around an industrial excavation in Boom Clay, Luxembourg, Luxembourg, 30–30 November 2005, Luxembuourg, Luxembourg; 2005. p. 305–9.

Bastiaens W, Bernier F. 25 years of underground engineering in a plastic clay formation: the HADES underground research laboratory. In: Bakker KJ, et al., editors. Geotechnical aspects of underground construction in soft ground. London: Taylor and Francis Group; 2006. p. 795–801.

Bastiaens W, Bernier F, Buyens M, et al. The connecting gallery-the extension of the HADES underground research facility at Mol, Belgium[J]. EURIDICE Rep. 2003;3:294.

Bastiaens W, Bernier F, Li XL. An overview of long-term HM measurements around HADES URF[C]. In: Proceedings of international symposium on multiphysics coupling and long-term behavior in rock mechanics, Liege; 2006. p. 15–26.

Bastiaens W, Bernier F, Li XL. SELFRAC: experiments and conclusions on fracturing, self-healing and self-sealing processes in clays. Phys Chem Earth. 2007;32(8–14):600–15.

Bayer JV, Jaeger F, Schaumann GE. Proton nuclear magnetic resonance (NMR) relaxometry in soil science applications. Open Magn Reson J. 2010;3:15–26.

Bear J, Corapcioglu CMY. A mathematical model for consolidation in a thermoelastic aquifer due to hot water injection or pumping. Water Resour Res. 1981;17(3):723–36.

Belmokhtar M, Delage P, Ghabezloo S, et al. Thermal volume changes and creep in the Callovo-Oxfordian claystone. Rock Mech Rock Eng. 2017;50:2297–309.

Bernabé Y, Mok U, Evans B. Permeability-porosity relationships in rocks subjected to various evolution processes. Pure Appl Geophys. 2003;160(5):937–60.

Bernier F, Neerdael B. Overview of in-situ thermomechanical experiments in clay: concept, results and interpretation. Eng Geol. 1996;41(1):51–64.

Bernier F, Volckaert G, Alonso E, Villar M. Suction controlled experiments on boom clay. Eng Geol. 1997;47(4):325–38.

Bernier F, Li XL, Verstricht J, Barnichon JD, Labiouse V, Bastiaens W, et al. CLIPEX: clay instrumentation programme for the extension of an underground research laboratory. Luxembourg: Commission of the European Communities; 2002. p. EUR20619.

Bernier F, Li XL, Bastiaens W. Twenty-five years' geotechnical observations and testing in the tertiary boom clay formation. Géotechnique. 2007;57(2):229–37.

F. Bernier, X. L. Li, W. Bastiaens, et al. Fractures and Self-healing within the Excavation Disturbed Zone in Clays (SELFRAC)- FINAL REPORT, EC PROJECT, EUR 22585, 2007.

Bernier F, Li XL, Bastiaens W, Ortiz L, Van Geet M, Wouters L, Frieg B, Blümling P, Desrues J, Viaggiani G, Coll C, Chanchole S, De Greef V, Hamza R, Malinsky L, Vervoort A, Vanbrabant Y, Debecker B, Verstraelen J, Govaerts A, Wevers M, Labiouse V, Escoffier S, Mathier JF, Gastaldo L, Bühler Ch. Fractures and Self-Healing within the Excavation Disturbed Zone in Clays (SELFRAC)- FINAL REPORT , EC PROJECT , EUR 22585, 2007.

Bésuelle P, Viggiani G, Desrues J, Coll C, Charrier P. A laboratory experimental study of the hydromechanical behavior of boom clay. Rock Mech Rock Eng. 2014;47:143–55.

Bird NRA, Preston AR, Randall EW, et al. Measurement of the size distribution of water-filled pores at different matric potentials by stray field nuclear magnetic resonance. Eur J Soil Sci. 2005;56(1):135–43.

Blüemling P, Bernier F, Lebon P, Martin CD. The excavation damaged zone in clay formations time-dependent behaviour and influence on performance assessment. Phys Chem Earth. 2007;32(8–14):588599.

Blumling P, Bernier F, Lebon P, Martin CD. The excavation damaged zone in clay formations time-dependent behaviour and influence on performance assessment. Phys Chem Earth. 2007;32:588–99.

Bonini M, Barla G, Barla M. Characterisation studies of Tectonised clay shales and implications in the excavation of large size tunnels[C]. In: ISRM 2003–technology roadmap for rock mechanics, Southern African Institute of Mining and Metallurgy; 2003.

Bossart P, Trick T, Meier PM, Mayor JC. Structural and hydrogeological characterisation of the excavation-disturbed zone in the Opalinus clay (Mont Terri project, Switzerland). Appl Clay Sci. 2004;26:429–48.
Bouyoucos GJ. Effect of temperature on some of the most important physical processes in soils[R]. East Lansing: Agricultural Experiment Station, Michigan State University; 1915.
Brett-Adams AC, Diamond LW, Petrini C, Madonna C. Influence of in-situ temperatures and pressures on the permeability of hydrothermally altered basalts in the oceanic crust. Tectonophysics. 2021;815:228994.
Bühler C, Heitz D, Trick T, Frieg B. In-situ self-sealing of the EDZ as a consequence of loading. In: Davies C, Bernier F, editors. Impact of the excavation disturbed or damaged zone (EDZ) on the performance of radioactive waste geological repositories. Proceedings of a European Commission CLUSTER Conference and Workshop Luxembourg, 3 to 5 November 2003. EUR 21028 EN; 2003. p. 281–6.
Burghignoli A, Desideri A, Miliziano S. Deformability of clays under non isothermal conditions. Rev Ital Geotech. 1992;4(92):227–36.
Burghignoli A, Desideri A, Miliziano S. A laboratory study on the thermomechanical behavior of clays. Can Geotech J. 2000;37(4):764–80.
Campanella RG, Mitchell JK. Influence of temperature variations on soil behavior. J. Soil Mech Found Div. 1968;94(SM3):709–34.
Carman PC. Fluid flow through granular beds. Trans Inst Chem Eng. 1937;15:150–66.
Carrier WD III. Goodbye, Hazen; hello, Kozeny–Carman. J Geotech Geoenviron. 2003;129:1054–6.
Cekerevac C. Thermal effects on the mechanical behaviour of saturated clays: an experimental and constitutive study. Lausanne, Switzerland: EPFL; 2003. (Ph.D. thesis), 258 pages
Cekerevac C, Laloui L. Experimental study of thermal effects on the mechanical behavior of a clay. Int J Numer Anal Methods Geomech. 2004;28(3):209–28.
Chaboche JL. Continuous damage mechanics—a tool to describe phenomena before crack initiation. Nucl Eng Des. 1981;64(2):233–47.
Charlier R, Chambon R, Collin F, et al. TIMODAZ report: deliverable D13——simulation of lab and in situ tests[R]. TIMODAZ; 2010.
Chen ZH, Xie Y, Sun SG, et al. Temperature-controlled triaxial apparatus for soils and its application[J]. Chinese J Geotech Eng. 2005;27(8):928–33. (in Chinese)
Chen GJ, Sillen X, Verstricht J, et al. ATLAS III in situ heating test in boom clay: field data, observation and interpretation. Comput Geotech. 2011;38(5):683–96.
Chen GJ, Maes T, Vandervoort F, Sillen X, Van Marcke P, Honty M, Vanderniepen P. Thermal impact on damaged boom clay and Opalinus clay: permeameter and isostatic tests with μCT scanning. Rock Mech Rock Eng. 2014;47(1):87–99.
Chen WZ, Gong Z, Yu HD, et al. Review of thermo-hydro-mechanical coupled tests and constitutive models of clay [J]. Rock Soil Mech. 2015;36(5):1217–36. (in Chinese)
Chen WZ, Ma YS, Yu HD, Li FF, Li XL, Sillen X. Effects of temperature and thermally-induced microstructure change on hydraulic conductivity of boom clay. J Rock Mech Geotech. 2017;9:383–95.
Chen WZ, Li FF, Ma YS, et al. Development of a parallel-linkage triaxial testing machine for THM coupling in soft rock[J]. Rock Soil Mech. 2019;40(03):1213–20. (in Chinese)
Chen GJ, Dizier A, Li XL, Verstricht J, Sillen X, Levasseur S. Numerical prediction of the large-scale in situ PRACLAY heater test in the boom clay. Rock Mech Rock Eng. 2021;54:2197–218.
Chen GJ, Li XL, Sillen X, Levasseur S. Determination of the anisotropic thermal conductivity of the boom clay based on a combined numerical interpretation of two in situ heater tests at different scales. Acta Geotech. 2022a;18(1):1–21.
Chen J, Tong H, Yuan J, Fang Y, Gu R. Permeability prediction model modified on Kozeny-Carman for building foundation of clay soil. Buildings. 2022b;12(11):1798.
Chen GJ, Yu L, Li XL, Levasseur S. Inverse analysis of two in situ heater tests in boom clay to identify a range of anisotropic thermal conductivities. Comput Geotech. 2024;165:105888.

Chijimatsu M, Fujita T, Sugita Y, et al. Field experiment, results and THM behavior in the Kamaishi mine experiment. Int J Rock Mech Min Sci. 2001;38(1):67–78.

Cho WJ, Lee JO, Chun KS. The temperature effects on hydraulic conductivity of compacted bentonite. Appl Clay Sci. 1999;14(1):47–58.

Coll C. Endommagement des Roches Argileuses et Perméabilité Induite au Voisinage d'Ouvrages Souterrains., PhD Thesis. Grenoble: Université Joseph Fourier; 2005.

Coll C, Escoffier S, Hamza RW, Vervoort A, Blumling P, Frieg B, Bastiaens W, Berest P, Bazargan B, Desrues J, Viggiani G, Labiouse V, Dehandschutter B, Wouters L, Vanbrabant Y, Mertens J, Li XL, Bernier F. SELFRAC: fractures and self-sealing within the excavation disturbed zone in clays—Deliverable 1, SELFRAC , EC project , 2004.

Costa A. Permeability-porosity relationship: a reexamination of the Kozeny-Carman equation based on a fractal pore-space geometry assumption. Geophys Res Lett. 2006;33(2):L02318.

Crisci E, Ferrari A, Giger SB, Laloui L. Anisotropic behaviour of shallow Opalinus clay. In: Ferrari A, Laloui L, editors. Energy geotechnics, SEG 2018, springer series in Geomechanics and geoengineering. Cham: Springer; 2019. https://doi.org/10.1007/978-3-319-99670-7_55.

Cui YJ, Ye WM. On modeling of thermo-mechanical volume change behavior of saturated clays[J]. Chin J Rock Mech Eng. 2006;24(21):3903–10. (in Chinese)

Cui YJ, Sultan N, Delage P. A thermomechanical model for saturated clays. Can Geotech J. 2000;37(3):607–20.

Cui YJ, Le TT, Tang AM, et al. Investigating the time-dependent behavior of boom clay under thermo-mechanical loading. Geotechnique. 2009;59(4):319–29.

Dafalias YF. Bounding surface elastoplasticity viscoplasticity for particulate cohesive media[C]. In: IUTAM symposium on deformation and failure of granular materials. Rotterdam: AA Balkema; 1982.

Datta R, Barr D, Boyle W. Measuring thermal, hydrological, mechanical, and chemical responses in the Yucca Mountain drift scale test. In: Coupled thermo-hydro-mechanical-chemical processes in geo-system[M]. Amsterdam: Elsevier; 2004.

David C, Wong TF, Zhu WL, Zhang JX. Laboratory measurement of compaction-induced permeability change in porous rocks: Implications for the generation and maintenance of pore pressure excess in the crust. Pure and applied geophysics. 1994;143(1–3):425–456.

Delage P. DISCUSSION: Compaction behaviour of clay. A. TARANTINO and E. DE COL. Géotechnique. 2009, 59(1), 75–77.

De Bruyn D, Labat S. The second phase of ATLAS: the continuation of a running THM test in the HADES underground research facility at Mol. Eng Geol. 2002;64(2):309–16.

De Bruyn D, Thimus JF. The influence of temperature on mechanical characteristics of boom clay: the results of an initial laboratory programme. Eng Geol. 1996;41(1):117–26.

De Bruyn D, Neerdael B, Gonze P, Rossion JP, Rousset G. Time dependent behaviour of the boom clay at great depth—an application to the construction of a waste disposal facility. Comput Geotech. 1987;3(1):3–20.

De Craen M, Wang L, Van Geet M, Moors H. Geochemistry of boom clay pore water at the Mol site. Status 2004. Edition, SCK•GEN. Scientific Report BLG-990, 2004, p. 179.

De la Vaissière R, Armand G, Talandier J. Gas and water flow in an excavation-induced fracture network around an underground drift: a case study for a radioactive waste repository in clay rock. J Hydrol. 2015;521:141–56.

Dehandschutter B, Vandycke S, Sintubin M, et al. Microfabric of fractured boom clay at depth: a case study of brittle–ductile transitional clay behavior. Appl Clay Sci. 2004;26(1):389–401.

Dehandschutter B, Vandycke S, Sintubin M, Vandenberghe N, Wouters L. Brittle fractures and ductile shear bands in argillaceous sediments: inferences from Oligocene boom clay (Belgium). J Struct Geol. 2005;27:1095–112.

Delage P, Sultan N, Cui YJ. On the thermal consolidation of boom clay. Can Geotech J. 2000;37(2):343–54.

Delage P, Cui YJ, Sultan N. On the thermal behaviour of boom clay. In: Proceedings Eurosafe 2004 conference, Berlin, Germany; 2004.

Delage P, Le TT, Tang AM, Cui YJ, Li XL. Suction effects in deep boom clay block samples. Géotechnique. 2007;57(1):239–44.

Delay J, Vinsot A, Krieguer JM, Rebours H, Armand G. Making of the underground scientific experimental programme at the Meuse/haute-Marne underground research laboratory, north eastern France. Phys Chem Earth. 2007;32:2–18.

Demars KR, Charles RD. Soil volume changes induced by temperature cycling. Can Geotech J. 1982;19:188–94.

Deng YF, Tang AM, Cui YJ, Nguyen XP, Li XL, Wouters L. Laboratory hydro-mechanical characterisation of boom clay at Essen and Mol. Phys Chem Earth. 2011;36:1878–90.

Derjaguin BV, Kasasev VV, Khromova EN. Thermal expansion of water in fine pores. Colloid Interface Sci. 1986;106:586–7.

Derski W, Kowalski ST. Equations of linear thermoconsolidation. Arch Mech. 1979;31(3):303–16.

Dewhurst DN, Yang Y, Aplin AC. Permeability and fluid flow in natural mudstones. Geol Soc Spec Pub. 1999;158:23–43.

Distinguin M, Lavanchy JM. Determination of hydraulic properties of the Callovo-Oxfordian argillite at the bure site: synthesis of the results obtained in deep boreholes using several in situ investigation techniques. Phys Chem Earth. 2007;32:379–92.

Dizier A, Chen G, Li X L, et al. The PRACLAY heater test after two years of the stationary phase[M]. EURIDICE Report, 2017.

Dizier A, Chen GJ, Verstricht J, Li XL, Sillen X, Levasseur S. The large-scale in situ PRACLAY heater test: first observations on the in situ thermo-hydro-mechanical behaviour of boom clay. Int J Rock Mech Min Sci. 2021;137:104558.

Djéran I, Bazargan B, Giraud A. Etude expérimentale du comportement thermo-hydro-mécanique de l'argile de Boom[R]. Brussels: ONDRAF; 1994.

Dolarevic S, Ibrahimbegovic A. A modified three-surface elasto-plastic cap model and its numerical implementation. Comput Struct. 2007;85(7):419–30.

Duan Z, Skoczylas F, Wang CR, Talandier J. Hydric cycle impacts on COx argillite permeability and young's modulus. Rock Mech Rock Eng. 2021;54:1129–47.

Fabre G, Pellet F. Creep and time-dependent damage in argillaceous rocks. Int J Rock Mech Min Sci. 2006;43:950–60.

Fox PJ, Edil TB. Effects of stress and temperature on secondary compression of peat. Can Geotech J. 1996;33(3):405–15.

François B, Laloui L, Laurent C. Thermo-hydro-mechanical simulation of ATLAS in situ large scale test in boom clay. Comput Geotech. 2009;36(4):626–40.

François B, Collin F, Dizier A, Charlier R. An extended Drucker-Prager hardening model for cross-anisotropy of soft rocks. In: Proceedings of the 15th European Conference on Soil Mechanics and Geotechnical Engineering; 2011. p. 537–40.

František H. Creep in soft soils. [Ph.D. Thesis] [D]. Trondheim: Norwegian University of Science and Technology; 2004.

Fujii T, Funatsu T, Oikawa Y, Sorai M, Lei X. Evolution of permeability during fracturing processes in rocks under conditions of geological storage of CO_2. Mater Trans. 2015;56(5):679–86.

Gasc-Barbier M, Chanchole S, Bèrest P. Creep behavior of bure clayey rock. Appl Clay Sci. 2004;26:449–58.

Gbewade CAF, Grgic D, Giraud A, et al. Experimental study of the effect of temperature on the mechanical properties of the Callovo-Oxfordian claystone. Rock Mech Rock Eng. 2024;57:4191–212.

Gens A, Garcia-Molina AJ, Olivella S, et al. Analysis of a full scale in situ test simulating repository conditions. Int J Numer Anal Methods Geomech. 1998;22(7):515–48.

Gens A, Vaunat J, Garitte B, Wileveau Y. In-situ behaviour of a stiff layered clay subject to thermal loading: observations and interpretation. Geotechnique. 2007;57(2):207–28.

Ghabezloo S, Sulem J, Saint-Marc J. Evaluation of a permeability-porosity relationship in a low-permeability creeping material using a single transient test. Int J Rock Mech Min Sci. 2009;46(4):761–8.

Ghahremannejad B. Thermo-mechanical behaviour of two reconstituted clays. Australia: University of Sydney; 2003. (Ph.D. thesis), 225 pages

Ghahremannejad B. Thermo-mechanical behavior of two reconstituted clays [D]. Sydney: Sydney University; 2006.

Giraud A, Rousset G. Time-dependent behavior of deep clays. Eng Geol. 1996;41(1):181–95.

Godefroy S, Korb JP, Fleury M, et al. Surface nuclear magnetic relaxation and dynamics of water and oil in macroporous media. Phys Rev E. 2001;64(2):021605.

Gong Z. Long-term thermo-hydro-mechanical coupled behavior of Belgium boom clay. Wuhan: Institute of Rock and Soil Mechanics, Chinese Academy of Sciences; 2015. [Ph.D. Thesis] [D] (in Chinese)

Gong Z, Chen WZ, Yu HD, et al. A thermo-mechanical coupled elastoplastic damage model for boom clay[J]. Rock Soil Mech. 2016;37(9):2433–42. (in Chinese)

Gonzalez-Blanco L, Romero E, Levasseur S. Self-sealing of boom clay after gas transport. Rock Mech Rock Eng. 2024;57:4173–89.

Graham J, Tanaka N, Crilly T, et al. Modified cam-clay modelling of temperature effects in clays. Can Geotech J. 2001;38(3):608–21.

Guo X, Zou GF, Wang YH, Wang Y, Gao T. Investigation of the temperature effect on rock permeability sensitivity. J Pet Sci Eng. 2017;156:616–22.

Hajiabdolmajid V, Kaiser PK, Martin CD. Modelling brittle failure of rock. Int J Rock Mech Min Sci. 2002;39(6):731–41.

Hamidi A, Tourchi S, Khazaei C. Thermomechanical constitutive model for saturated clays based on critical state theory. Int J Geomech. 2015;15(1):4014038.

Hamidi A, Tourchi S, Kardooni F. A critical state based thermo-elasto-plastic constitutive model for structured clays. J Rock Mech Geotech Eng. 2017;9(6):1094–103.

Harrington JF, Noy DJ, Cuss RJ. Callovo oxfordian claystone: processes governing advective gas flow, British Geological Survey, Minerals and Waste Programme, Commissioned Report CR/13/088, 2013. EU REPORT D5.16.

Harrington JF, Volckaert G, Noy DJ. Long term impact of temperature on the hydraulic permeability of bentonite. In: Norris S, Bruno J, Cathelineau M, Delage P, Fairhurst C, Gaucher EC, Höhn EH, Kalinichev A, Lalieux P, Sellin P, editors. Clays in natural and engineered barriers for radioactive waste confinement. London: Geological Society; 2014. p. 589–601. Special Publications, 400.

Holtz RD, Kovacs WD. An introduction to geotechnical engineering, vol. 733. Englewood Cliffs, N. J.: Prentice-Hall; 1981.

Hong PY, Pereira JM, Tang AM, Cui YJ. On some advanced thermo-mechanical models for saturated clays. Int J Num Anal Meth Geomech. 2013;37(17):2952–71.

Horseman ST, Winter MG, Entwistle DC. Geotechnical characterization of boom clay in relation to the disposal of radioactive waste, Commission of the European Communities, Report EUR; 1987. p. 10987.

Horseman ST, Winter MG, Entwistle DC. Triaxial experiments on boom clay. In: Cripps JC, Coulthard JM, Culshaw MG, Forster A, Hencher SR, Moon CF, editors. The engineering geology of weak rock. Rotterdam: Balkema; 1993. p. 36–43.

Hou F, Li Q, Liu E, Zhou C, Liao M, Luo H, et al. A fractional creep constitutive model for frozen soil in consideration of the strengthening and weakening effects. Adv Mater Sci Eng. 2016;2016:1.

Houben ME, Van Eeden JCM, Barnhoorn A, Hangx SJT. Fracture induced permeability in Whitby mudstone. Environ Sci Technol. 2020;54(15):9564–72.

Houhou R, Sutman M, Sadek S, Laloui L. Microstructure observations in compacted clays subjected to thermal loading. Eng Geol. 2021;287:105928.

Hoxha D, Jiang Z, Homand F, et al. Impact of THM constitutive behavior on the rock-mass response: case of HE-D experiment in Mont-Terri underground rock laboratory[C]. In: EUROCK 2006: Multiphysics coupling and long term behavior in rock mechanics. Liège: Balkema; 2006.

Hu HJ, Braun P, Delage P, Ghabezloo S. Evaluation of anisotropic poroelastic properties and permeability of the Opalinus clay using a single transient experiment. Acta Geotech. 2021;16:2131–42.
Hueckel T. Chemo-plasticity of clays subjected to stress and flow of a single contaminant. Int J Numer Anal Meth Geomech. 1997;21(1):43–72.
Hueckel T, Baldi G. Thermoplasticity of saturated clays: experimental constitutive study. J Geotech Eng. 1990;116(12):1778–96.
Hueckel T, Borsetto M. Thermoplasticity of saturated soils and shales: constitutive equations. J Geotech Eng. 1990;116(12):1765–77.
Hueckel T, Pellegrini R. Modelling of thermal failure of saturated clays. In: Proceedings of the 3rd symposium numerical modeling in geomechanics. London: Elsevier; 1989. p. 81–90.
Hueckel T, Pellegrini R. Thermoplastic modeling of undrained failure of saturated clay due to heating. Soils Found. 1991;31(3):1–16.
Hueckel T, Pellegrini R. Effective stress and water pressure in saturated clays during heating-cooling cycles. Can Geotech J. 1992;29(6):1095–102.
Hueckel T, Pellegrini R, Olmo CD. A constitutive study of thermo-elasto-plasticity of carbonatic clays. Int J Numer Anal Methods Geomech. 1998;22:549–74.
Hueckel T, Francois B, Laloui L. Explaining thermal failure in saturated clays. Géotechnique. 2009;59(3):197–212.
Huysmans M, Berckmans A, Dassargues A. Effect of excavation induced fractures on radionuclide migration through the boom clay (Belgium). Appl Clay Sci. 2006;33(3):207–18.
Ishizaki T, Maruyama M, Furukawa Y, et al. Premelting of ice in porous silica glass. J Cryst Growth. 1996;163(4):455–60.
Jefferson I. Temperature effects on clay soils. [Ph.D. Thesis][D]. Loughborough University; 1994.
Jia SP. Hydro-mechanical coupled creep damage constitutive model of boom clay, back analysis of model parameters and its engineering application. Wuhan: Institute of Rock and Soil Mechanics, Chinese Academy of Sciences; 2009. [Ph.D. Thesis] [D] (in Chinese)
Jia SP, Chen WZ, Yu HD, et al. Study of hydro-mechanical-damage coupled creep constitutive model of mudstone, part I: theoretical model[J]. Rock Soil Mech, 2011, 32(09): 2596–2602. (in Chinese).
Jivkov AP, Hollis C, Etiese F, McDonald SA, Withers PJ. A novel architecture for pore network modelling with applications to permeability of porous media. J Hydrol. 2013;486:246–58.
Joshaghani M, Ghasemi-Fare O. Exploring the effects of temperature on intrinsic permeability and void ratio alteration through temperature-controlled experiments. Eng Geol. 2021;293:106299.
Knowles NC, Jeffries RM, Come B. Benchmarking the thermomechanical behavior of clays—a progress report—the CEC interclay project. Eng Geol. 1996;41(1):65–71.
Kontogianni V, Psimoulis P, Stiros S. What is the contribution of time-dependent deformation in tunnel convergence. Eng Geol. 2006;82:264–7.
Kuntiwattanakul P, Towhata I, Ohishi K, et al. Temperature effects on undrained shear characteristics of clay. Soils Found. 1995;35(1):147–62.
Kurz D, Sharma J, Alfaro M, Graham J. Semi-empirical elastic–thermoviscoplastic model for clay. Can Geotech J. 2016;53(10):1583–99.
Kwon O, Kronenberg AK, Gangi AF, Johnson B. Permeability of Wilcox shale and its effective pressure law. J Geophys Res-Solid Earth. 2001;106:19339–53.
Labiouse V. Advanced hydro-mechanical modelling of excavations in the Boom Clay formation. In: Extended abstracts of the Boom clay seminar, Alden Biesen; 1997. p. 45–61.
Laloui L, Cekerevac C. Thermo-plasticity of clays: anisotropic yield mechanism. Comput Geotech. 2003;30(8):649–60.
Laloui L, François B. Acmeg-T: soil thermoplasticity model. J Eng Mech. 2009;135(9):932–44.
Lambe TW. A mechanical picture of shear strength in clay [C]. In: Research conference on shear strength of cohesive soils. Boulder, CO: University of Colorado; 1960. p. 555–80.
Le TT. Comportement thermo-hydro-mécanique de l'argile de Boom. PhD thesis. Paris: CERMES, Ecole Nationale des Ponts et Chaussées; 2008.

Lebon P, Ghoreychi M. French underground research Laboratory of Meuse/haute-Marne: THM aspects of argillite formation[R]. EUROCK2000, Aachen, 2000.
Li CG. Void ratio-permeability relations for clays. Quarterly Journal of Engineering Geology and Hydrogeology. 2020;53(2):298–303.
Li XL, Bastiaens W, Van Marcke P, et al. Design and development of large-scale in-situ PRACLAY heater test and horizontal high-level radioactive waste disposal gallery seal test in Belgian HADES. J Rock Mech Geotech Eng. 2010;2(2):103–10.
Li XL, Neerdael B, Raymaekers D, Sillen X. The construction of the HADES underground research laboratory and its role in the development of the Belgian concept of a deep geological repository. Geol Soc Lond, Spec Publ. 2023;536(1):159–84.
Li TB, Peng F, Chen C, Ma CC, Li YL, Wang YX, Li YS. Investigating the creep damage and permeability evolution mechanism of Phyllite considering non-Darcy' s flow. Rock Mech Rock Eng. 2024;57:8483–98.
Lima A. Thermo-hydro-Mechanical Behaviour of Two Deep Belgian Clay Formations: Boom and Ypresian Clays [Ph.D. thesis]. Spain: Universitat Politècnica de Catalunya; 2011.
Litewka A, Debinski J. Load-induced oriented damage and anisotropy of rock-like materials. Int J Plast. 2003;19(12):2171–91.
Liu EL, Xing HL. A double hardening thermo-mechanical constitutive model for overconsolidated clays. Acta Geotech. 2009;4(1):1–6.
Liu XJ, Gao H, Lian LX. Study of temperature and confining pressure effects on porosity and permeability in low permeability sandstone. Chin J Rock Mech Eng. 2011;30(S2):3771–8.
Liu Z, Shao J, Xie S, Conil N, Talandier J. Mechanical behavior of claystone in lateral decompression test and thermal effect. Rock Mech Rock Eng. 2019;52(2):321–34.
Lyakhovsky V, Hamiel Y. Damage evolution and fluid flow in poroelastic rock. Izv Phys Solid Earth. 2007;43(1):13–23.
Ma YS, Chen WZ, Yu HD, Gong Z, Li XL. Variation of the hydraulic conductivity of boom clay under various thermal-hydro-mechanical conditions. Eng Geol. 2016;212:35–43.
Mahnken R. Theoretical, numerical and identification aspects of a new model class for ductile damage. Int J Plast. 2002;18(7):801–31.
Mánica M, Gens A, Vaunat J, Ruiz DF. A time-dependent anisotropic model for argillaceous rocks. Application to an underground excavation in Callovo-Oxfordian claystone. Computers and Geotechnics. 2017, 85: 341–50.
Menaceur H, Delage P, Tang A-M, Conil N. The thermomechanical behaviour of the Callovo-Oxfordian claystone. Int J Rock Mech Min Sci. 2015;78:290–303.
Menaceur H, Delage P, Tang AM, Conil N. On the thermo-hydro-mechanical behaviour of a sheared callovo-oxfordian claystone sample with respect to the EDZ behaviour. Rock Mech Rock Eng. 2016;49(5):1875–88.
Mertens J, Bastiaens W, Dehandschutter B. Characterisation of induced discontinuities in the boom clay around the underground excavations (URF, Mol, Belgium). Appl Clay Sci. 2004;26:413–28.
Meschyan SR, Galstyan RR. Investigation of compressional creep of soil with consideration of temperature effects. Soil Mech. Found. Eng. 1972;9(4):227–31.
Mitchell JK, Soga K. Fundamentals of soil behavior[M]. New York: Wiley; 1976.
Modaressi H, Laloui L. A thermo-viscoplastic constitutive model for clays [J]. Int J Numer Anal Methods Geomech. 1997;21(5):313–35.
Mohajerani M, Delage P, Sulem J, Monfared M, Tang A-M, Gatmiri B. The thermal volume changes of the Callovo-Oxfordian claystone. Rock Mech Rock Eng. 2013;47:131–42.
Monfared M, Delage P, Sulem J, Mohajerani M, Tang AM, De Laure E. A new hollow cylinder triaxial cell to study the behavior of geo-materials with low permeability. Int J Rock Mech Min. 2011a;48:637–49.
Monfared M, Sulem J, Delage P, Mohajerani M. A laboratory investigation on thermal properties of the Opalinus claystone. Rock Mech Rock Eng. 2011b;44:735–47.

Monfared M, Sulem J, Delage P, Mohajerani M. On the THM behaviour of a sheared boom clay sample: application to the behaviour and sealing properties of the EDZ. Eng Geol. 2012;124:47–58.
Monfared M, Sulem J, Delage P, Mohajerani M. Temperature and damage impact on the permeability of Opalinus clay. Rock Mech Rock Eng. 2014;47:101–10.
Morin R, Silva AJ. The effects of high pressure and high temperature on some physical properties of ocean sediments. J Geophys Res: Solid Earth (1978–2012). 1984;89(B1):511–26.
Murakami S. Creep damage and creep-fatigue damage. In: Continuum damage mechanics—a continuum mechanics approach to the analysis of damage and fracture. New York: Springer; 2012. p. 217–51.
Nagaraj TS, Pandian NS, Raju PSRN. Stress-state-permeability relations for overconsolidated clays. Geotechnique. 1994;44(2):349–52.
Naumann M, Hunsche U, Schulze O. Experimental investigations on anisotropy in dilatancy, failure and creep of Opalinus clay. Phys Chem Earth. 2007;32:889–95.
Neerdael B, Beaufays R, Buyens M, De Bruyn D, Voet M. Geomechanical behaviour of boom clay under ambient and elevated temperature conditions. In: Final report Commission of the European Communities Nuclear Science and Technology EUR 14154; 1992.
Noble CA. Effect of temperature on strength of soils [D]. Ames: Iowa State University; 1968.
Nowak T, Kunz H, Dixon D, et al. Coupled 3-D thermo-hydro-mechanical analysis of geotechnological in situ tests. Int J Rock Mech Min Sci. 2011;48(1):1–15.
Nüesch R. Das mechanische Verhalten von Opalinuston. (Dissertation). Switzerland: ETH Zürich; 1991.
Paaswell RE. Temperature effects on clay soil consolidation. J Soil Mech Found Div ASCE. 1967;93(SM3):9–22.
Paterson MS. The equivalent channel model for permeability and resistivity in fluid saturated rock: a re-appraisal. Mech Mater. 1983;2(4):345–52.
Perzyna P. Fundamental problems in viscoplasticity. Adv Appl Mech. 1966;9:243.
Perzyna P. Thermodynamic theory of viscoplasticity. Adv Appl Mech. 1971;11:313–54.
Picard JM, Bazargan B, Rousset G. Essai thermo-hydro-mechanique dans une argile profonde: Essai CACTUS. CEC report. EUR 15482 FR. Luxembourg: CEC; 1994.
Piriyakul K, Haegeman W. Stiffness anisotropy of boom clay. In: Proceedings of the 17th International Conference on Soil Mechanics and Geotechnical Engineering, 5–9 October 2009, Alexandria, Egypt, vol. 1; 2009. p. 167–71.
Plum RL, Esrig MI. Some temperature effects on soil compressibility and pore water pressure[R], vol. 103. Washington D.C.: Highway Research Board., Special Report; 1969. p. 231–42.
Pons CH, Franzetti N, Tchoubar D. Analyse de la structure multiéchelle des dispersions de montmorillonite Na. CR Rech. 1994; 92 N 800094, PIRSEM et ARTEP
Popp T, Salzer K. Salzer anisotropy of seismic and mechanical properties of Opalinus clay during triaxial deformation in a multi-anvil apparatus. Phys Chem Earth. 2007;32(8–14):879–88.
Pusch R, Güven N. Electron microscopic examination of hydrothermally treated bentonite clay. Eng Geol. 1990;28(3):303–14.
Qiao YF; Ding WQ. ACMEG-TVP: A thermoviscoplastic constitutive model for geomaterials. Computers and Geotechnics. 2017;81:98–111.
RAC. Fractures and self-healing within the excavation disturbed zone in clays—Deliverable 4; 2004.
Raoof A, Nick HM, Hassanizadeh SM, Spiers CJ. Pore flow: a complex pore-network model for simulation of reactive transport in variably saturated porous media. Comput Geosci. 2013;61:160–74.
Ren J, Shen ZZ, Yang J, Zhao J, Yin JN. Effects of temperature and dry density on hydraulic conductivity of silty clay under infiltration of low-temperature water. Arab J Sci EngGeol. 2014;39:461–6.
Robinet JC, Rahbaoui A, Plas F, et al. A constitutive thermomechanical model for saturated clays. Eng Geol. 1996;41(1):145–69.

Romero E, Gens A, Lloret A. Temperature effects on the hydraulic behavior of an unsaturated clay. Geotech Geol Eng. 2001;19(3–4):311–32.

Rousset G. Comportement mécanique des argiles profondes. Application au stockage des déchets radioactifs. Unpublished PhD thesis, Ec. Net. Ponts & chaussees, 1988.

Rutqvist J, Tsang CF. Analysis of thermal–hydrologic–mechanical behavior near an emplacement drift at Yucca Mountain. J Contam Hydrol. 2003;62:637–52.

Rutqvist J, Börgesson L, Chijimatsu M, et al. Coupled thermo-hydro-mechanical analysis of a heater test in fractured rock and bentonite at Kamaishi mine—comparison of field results to predictions of four finite element codes. Int J Rock Mech Min Sci. 2001;38(1):129–42.

Rutqvist J, Backstrom A, Chijimatsu M, Feng XT, Pan PZ, Hudson J, et al. A multiple-code simulation study of the long-term EDZ evolution of geological nuclear waste repositories. Environ Geol. 2009;57:1313–24.

Schiffmann RL. A thermoelastic theory of consolidation. In: Cremen ET, et al., editors. Environmental and geophysical heat transfer. New York, NY: American Soc. of Mech. Engrs; 1971.

Schmertmann JH, Osterberg JO. An experimental study of the development of cohesion and friction with axial strain in saturated cohesive soils[C]. In: Research conference on shear strength of cohesive soils. ASCE; 1960. p. 643–94.

Schuster V, Rybacki E, Bonnelye A, et al. Experimental deformation of Opalinus clay at elevated temperature and pressure conditions: mechanical properties and the influence of rock fabric. Rock Mech Rock Eng. 2021;54:4009–39.

Shao H, Sönnke J, Morel J, Krug S. In situ determination of anisotropic permeability of clay. Phys Chem Earth. 2011;36:1688–92.

Shao ZS, Wu K, Yuan Y. Theoretical investigation on rock rheological characteristics in variable temperature field [J]. Chinese J Appl Mech. 2018;35(05):962–968+1180. (in Chinese)

Shi X, Yang X, Yue W, Yang L, Gui J, Huang H, Jiang S. Experimental investigation on the creep behaviors of shale using nanoindentation technique and fractional constitutive models. J Pet Sci Eng. 2022;215(Part A):110520.

Sojoudi M, Li B. A thermodynamic-based model for modeling thermo-elastoplastic behaviors of saturated clayey soils considering bound water dehydration. J Rock Mech Geotech Eng. 2023;15:1535–46.

Song F, Gens A, Collico S, Grgic D, Wang H, Fully coupled THM constitutive model for clay rocks: Formulation and application to laboratory tests, Journal of Rock Mechanics and Geotechnical Engineering, 2025, 17(4): 1941–60.

Sultan N. Etude du comportement thermo-mécanique de l'argile de Boom: expériences et modélisation. PhD thesis. Ecole Nationale des Ponts et Chaussées; 1997. p. 217.

Sultan N, Delage P, Cui YJ. Temperature effects on the volume change behaviour of boom clay. Eng Geol. 2002;64:135–45.

Tanaka N, Graham J, Crilly T. Stress-strain behavior of reconstituted illitic clay at different temperatures. Eng Geol. 1997;47(4):339–50.

Tang CA, Tham LG, Lee PKK, Yang TH, Li LC. Coupled analysis of flow, stress and damage (FSD) in rock failure. Int J Rock Mech Min Sci. 2002;39(4):477–89.

Tang AM, Cui YJ, Tali B, Doan DH. Etude du comportement thermo-hydro-mécanique visqueux de l'argile de Boom[R]. Paris: École des Ponts Paris Tech; 2008.

Taylor D. Fundamentals of soil mechanics. New York: Wiley; 1948. p. 66.

Terzaghi K. Theoretical soil mechanics. New York: Wiley; 1943.

Thomas F, Jin HM, Cases JM. Définition du comportement des argiles au cours des forages. C.R. Rech. 1994; 92 N 80003, PIRSEM et ARTEP

Tian HH, Wei CF. A NMR-based testing and analysis of adsorbed water content. Sci Sin Technol. 2014;44(3):295–305.

Tidfors M, Sallfors G. Temperature effect on preconsolidation pressure. Geotech Test J. 1989;01:93–7.

Towhata I, Kuntiwattanakul P, Seko I, Ohishi K. Volume change of clays induced by heating as observed in consolidation tests. Soils Found. 1993a;33(4):170–83.
Towhata I, Kuntiwattanakul P, Kobayashi H. A preliminary study on heating of clays to examine possible effects of temperature on soil-mechanical properties. Soils Found. 1993b;33(4):184–90.
Tribaudino M, Angel RJ, Cámara F, Nestola F, Pasqual D, Margiolaki I. Thermal expansion of plagioclase feldspars. Contrib Mineral Petrol. 2010;160:899–908.
Tsai LS, Hsieh YM, Weng MC, Huang TH, Jeng FS. Time-dependent deformation behaviors of weak sandstones. Int J Rock Mech Min Sci. 2008;45:144–54.
Van Geet M, Bastiaens W, Ortiz L. Self sealing capacity of argillaceous rocks: review of laboratory results obtained from the SELFRAC project. Phys Chem Earth. 2008;33:S396–406.
Van Marcke P, Bastiaens W. Excavation induced fractures in a plastic clay formation: observations at the HADES URF. J Struct Geol. 2010;32(11):1677–84.
Van Marcke P., Li X., Bastiaens W., Verstricht J., Chen G., Leysen J., Rypens J. The design and installation of the PRACLAY In-Situ Experiment. EURIDICE Report 13-129, Mol, Belgium, 2013, 190 p.
Vervoort A, Bastiaens W, Bernier F, et al. Fractures and self-healing within the excavation disturbed zone in clays. Laboratory tests (results and interpretation). Deliverable 2, SELFRAC project, EC project, 2005.
Voltolini M, Ajo-Franklin JB. The sealing mechanisms of a fracture in opalinus clay as revealed by in situ synchrotron x-ray micro-tomography. Front Earth Sci. 2020;16:1–13.
Wang SW, Zhu W, Qian XD, Xu HQ, Fan XH. Temperature effects on non-Darcy flow of compacted clay. Applied Clay Science. 2017;135:521–525.
Wang Y, Shi B, Gao L, et al. Laboratory tests for temperature effects of clay permeability[J]. J Eng Geol. 2010;18(3):351–6. (in Chinese)
Wang KJ, Shan ZG, Shen KM, Wang MY, Du WB, Mao BY. The thermo-mechanical behaviour of clay in different stress and temperature paths. In: Steyn WJ, Wang Z, Holleran G, editors. Transportation infrastructure engineering, materials, behavior and performance, GeoChina 2021. Sustainable civil infrastructures. Cham: Springer; 2021.
Weaver CE. Geothermal alteration of clay minerals and shales: diagenesis[R]. Atlanta: Georgia Institute of Technology; 1979.
Wikipedia. Double layer (interfacial). Wikipedia Website (2015) [EB]. http://en.wikipedia.org/wiki/Double_layer_(interfacial).
Yang Q. Study on creep mechanical properties of artificial frozen sandy mudstone in the cretaceous strata of western China[D]. Xi'an University of Science and Technology; 2018. (in Chinese)
Yang Y, Aplin AC. A permeability-porosity relationship for mudstones. Mar Pet Geol. 2010;27(8):1692–7.
Yang JP, Chen WZ, Tian HM, Yu HD. Study of permeability evolutions in low permeability medium under different stresses and temperatures. Rock Soil Mech. 2009;30(12):3587–95.
Yao YP, Yang YF, Niu L. UH model considering temperature effects[J]. Science China Technol Sci. 2011a;41(2):158–69.
Yao YP, Wan Z, Yang YF, et al. Thermal failure for saturated clay under undrained condition[J]. Rock Soil Mech. 2011b;32(9):2561–9. (in Chinese)
Yavari N, Tang AM, Pereira JM, Hassen G. Effect of temperature on the shear strength of soils and the soil-structure interface. Can Geotech J. 2016;53(7):1186–94.
Ye WM, Wan M, Chen B, Chen YG, Cui YJ, Wang J. Temperature effects on the swelling pressure and saturated hydraulic conductivity of the compacted GMZ01 bentonite. Environ Earth Sci. 2013;68(1):281–8.
Yong RN. Soil properties and behavior[M]. Amsterdam: Elsevier; 1975.
You LJ, Kang YL. Effects of thermal treatment on physical property of tight rocks. Prog Geophys. 2009;24(5):1850–4.
Yu HD. Study on long term hydro-mechanical coupled behavior of Belgium boom clay. Wuhan: Institute of Rock and Soil Mechanics, Chinese Academy of Sciences; 2010. [Ph.D. Thesis] [D]. (in Chinese)

Yu HD, Chen WZ, Jia SP, Cao JJ, Li XL. Experimental study on the hydro-mechanical behavior of boom clay. Int J Rock Mech Mining Sci. 2012;53:159–65.
Yu L, Rogiers B, Gedeon M, Marivoet J, De Craen M, Mallants D. A critical review of laboratory and in-situ hydraulic conductivity measurements for the boom clay in Belgium. Appl Clay Sci. 2013;75–76:1–12.
Yu HD, Chen WZ, Li XL, Sillen X. A transversely isotropic damage model for boom clay. Rock Mech Rock Eng. 2014;47:207–19.
Yu HD, Chen WZ, Gong Z, Tan XJ, Ma YS, Li XL, Sillen X. Creep behavior of boom clay. Int J Rock Mech Min Sci. 2015;76:256–64.
Yu HD, Chen WZ, Gong Z, Ma YS, Chen GJ, Li XL. Influence of temperature on the hydro-mechanical behavior of Boom clay. Int J Rock Mech Min Sci. 2018;108:189–97.
Yu HD, Lu C, Chen WZ, Tian HM. An insight into the creep mechanisms of a clayey soil through long-term consolidation tests. Bull Eng Geol Environ. 2021;80:9127–39.
Yu HD, Chen WZ, Gong Z, Tan XJ, Yang DS. Loading-unloading behavior of a clayey rock under thermo-hydro-mechanical conditions. Int J Rock Mech Min Sci. 2021a;148:104966.
Yu HD, Lu C, Chen WZ, Li HH. Permeability changes in fractured Tamusu mudstone in the context of radioactive waste disposal. Bull Eng Geol Environ. 2021b;80:7945–57.
Yu HD, Lu C, Chen WZ, Tian HM. An insight into the creep mechanisms of a clayey soil through long-term consolidation tests. Bull Eng Geol Environ. 2021c;80:9127–39.
Yu HD, Chen WZ, Ma YS, Gong Z, Tian HM, Yuan JQ. Thermo-hydro-mechanical behavior of a clayey rock: a constitutive approach and numerical validation. Case Studies Thermal Eng. 2022a;39:102424.
Yu HD, Chen WZ, Ma YS, Tan XJ, Yang JP. Experimental and theoretical study on the creep behavior of a clayey rock. Rock Mech Rock Eng. 2022b;56:1387–98.
Yu HD, Lu C, Chen WZ, Yang DS, Li HH. Thermo-mechanical effect on the anisotropic permeability of a mudstone. Case Studies Therm Eng. 2023a;41:102625.
Yu HD, Chen WZ, Ma YS, Gong Z, Lu HF. Thermo-mechanical effect on the drained creep behavior of a clay rock. Eng Geol. 2023b;327(4):107327.
Yu HD, Chen WZ, Ma YS, Tan XJ, Yang JP. Experimental and theoretical study on the creep behavior of a clayey rock. Rock Mech Rock Eng. 2023c;56:1387–98.
Yu HD, Lu C, Chen WZ. Mechanical behavior of Tamusu mudstone under elevated temperatures. Case Studies Therm Eng. 2024a;55:104174.
Yu HD, Chen WZ, Li FF, Tan XJ, Yang JP. Probing the hydro-mechanical response of a clayey rock for radioactive waste disposal using a transversely isotropic damage constitutive model. Rock Mech Rock Eng. 2024b;57:131–43.
Zhang YJ. Coupled thermo-hydro-mechanical model and 2D FEM analysis for FEBEX in-situ test[J]. Chinese J Geotech Eng. 2007;29(3):313–8. (in Chinese)
Zhang CL. Experimental evidence for self-sealing of fractures in claystone. Phys Chem Earth. 2011;36:1972–80.
Zhang CL. Sealing of fractures in claystone. J Rock Mech Geotech. 2013;5:214–20.
Zhang CL. The stress–strain–permeability behaviour of clay rock during damage and recompaction. J Rock Mech Geotech Eng. 2016;8(1):16–26.
Zhang ZC, Cheng XH. A thermo-mechanical coupled constitutive model for clay based on extended granular solid hydrodynamics. Comput Geotech. 2016;80(12):373–82.
Zhang ZC, Cheng XH. A fully coupled THM model based on a non-equilibrium thermodynamic approach and its application. IntJ Numer Anal Methods Geomech. 2017;41(4):527–54.
Zhang CL, Laurich B. Mechanical behavior of sandy facies of Opalinus clay under different load conditions. J Rock Mech Geotech. 2020;12:223–41.
Zhang CL, Rothfuchs T. Experimental study of the hydro-mechanical behaviour of the Callovo-Oxfordian argillite. Appl Clay Sci. 2004;26(1–4):325–36.
Zhang CL, Rothfuchs T. Damage and sealing of clay rocks detected by measurements of gas permeability. Phys Chem Earth. 2008;33:S363–73.

Zhang CL, Rothfuchs T, Su K. Experimental study of the thermo-hydro-mechanical behavior of indurated clays. Phys Chem Earth Parts A/B/C. 2007;32(8):957–65.
Zhang S, Leng W, Zhang F, et al. A simple thermo-elastoplastic model for geomaterials. Int J Plast. 2012;34:93–113.
Zhang CL, Conil N, Armand G. Thermal effects on clay rocks for deep disposal of high-level radioactive waste. J Rock Mech Geotech. 2017;9:463–78.
Zhou H, Jia Y, Shao JF. A unified elastic-plastic and viscoplastic damage model for quasi-brittle rocks. Int J Rock Mech Min Sci. 2008;45(8):1237–51.
Zhou H, Hu D, Zhang F, et al. A thermo-plastic/viscoplastic damage model for geomaterials. Acta Mech Solida Sin. 2011;24(3):195–208.
Zhu W, Montési L, Wong TF. Characterizing the permeability-porosity relationship during compactive cataclastic flow. In: Proceedings of the 42nd US rock mechanics and the 2nd US-Canada rock mechanics symposium. Madison, Wisconsin, US: Omni Press; 2008. p. 29–33.
Zihms SG, Harrington JF. Thermal cycling: impact on bentonite permeability. Mineral Mag. 2015;79(6):1543–50.
Zymnis DM, Whittle AJ. Geotechnical considerations in the design of borehole heat exchangers. Can Geotech J. 2020;58(9):1247–62.
Zymnis DM, Whittle AJ, Germaine JT. Measurements of temperature-dependent bound water in clays. ASTM Geotech Test J. 2018;42(1):232–44.
Zymnis DM, Whittle AJ, Cheng X. Simulation of long-term Thermo-mechanical response of clay using an advanced constitutive model. Acta Geotech. 2019;14:295–311.

MIX
Papier aus verantwortungsvollen Quellen
Paper from responsible sources
FSC® C105338

If you have any concerns about our products,
you can contact us on
ProductSafety@springernature.com

In case Publisher is established outside the EU,
the EU authorized representative is:
Springer Nature Customer Service Center GmbH
Europaplatz 3, 69115 Heidelberg, Germany

Printed by Libri Plureos GmbH
in Hamburg, Germany